Pandia

R.E.A.L. SCIEN

- Read
- Explore
- Absorb
- Learn

Life (level one)

For Grades 1 – 4

Written & Illustrated by
Terri Williams

www.PandiaPress.com

copyright Pandia Press
ISBN: 978-0-9766057-0-6

Pandia Press

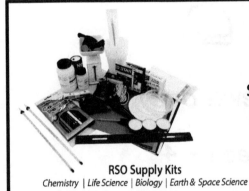

www.PandiaPress.com

WHAT'S IN THIS BOOK?

•Denotes the "For My Notebook" page *Denotes a lab or activity

AN INTRODUCTION TO R.E.A.L. Science Odyssey

This book was intended to be used from start to finish, much like a math book, and as such, vocabulary and concepts build on one another. You may find words and concepts in here that you feel the need to "review and practice." Feel free to do that if you wish but understand that vocabulary words are repeated throughout the story sections so your children will hear the same words many times. This is intended to help them learn without having to drill. Having said that, review can be good and anytime you can use a concept to refer to something you see in real life, your child will benefit greatly. Eating fish for dinner (fish sticks excluded)? How about showing your child the gills under those gill flaps? Taking a walk in the country? Point out the different footprints you see in the mud. Are they bird, reptile or amphibian? Next time you eat an egg talk about the cell that it is. Use real words for body parts. Babies grow up in a uterus, not a tummy. Science is only a foreign language if it isn't used in real life.

Used two or three times a week this book is a complete, rigorous, vocabulary-rich life science curriculum that needs no supplementation. It is not a collection of labs to do randomly and with no flow from one to the other nor is it an overwhelmingly long progressions of trivial facts to be memorized and forgotten (like we grew up with). This book was designed with the non-science parent in mind, so you could pick it up and follow along with no need for further organizing or research. This is the story of life, beginning with the child's own body for reference. This book is a minds-on and hands-on program. If you hate to touch "bugs," wouldn't think about letting your child handle worms, and have no intention of getting gooey and dirty, RUN NOW! Science is about experimentation, but there are read and report science curricula for the faint of heart. Are you still here? If so, roll up those sleeves and get ready to turn every household container into a critter keeper.

For every notebook page in this book, children will do 1-4 activities that build upon and reinforce what they have heard. Labs also teach new material, so it is important to try to do all of the labs included. In addition, we have included journaling ideas plus book and website suggestions for a complete indoor and outdoor experience and to dig into whatever your child finds most fascinating. You will notice that many labs are infused with grammar-age appropriate math. Science is very mathematical with measuring, graphing and calculating. If your child struggles with the math or with writing the results, don't let the lab papers overwhelm the lesson. The idea is to enjoy science. Much of the learning comes from doing and discussing. Read the questions to the prewriting child and have him dictate the answers back to you, or if you both find them tedious, skip them altogether.

A few words about big words. You know the ones I mean: monocot, chloroplast, echinoderm -EEEK! It's enough to make a person drop their science book and run in fear. Now, let me share a few more with you. How about Tyrannosaurus Rex, Triceratops and Velociraptor? All very long words that we have learned right along with our 3 and 4 year olds. Show no fear in the face of these latin based words and you too will be referring to arachnids at your next park day!

GETTING STARTED

1. If you are a computer person, pull out the website list (pgs. 17-18) and leave it by your computer so you can look things up as your child shows interest.
2. If you are a book person, pull out the library book list (pgs. 13-16) and go over it a few weeks before you do any given section. Leave it in your library book bag or by your computer so you can check on the availability of the books.

3. If you are a nervous "Did I teach my child enough stuff" type of person, go over the "What's the Big Idea" pages in this book. It will tell you what sorts of things your child should come out of this course knowing. Use it with a grain of salt. Remember, the key here is exposure and fun.
4. Look ahead to what material you will need for the upcoming week or year if you're an uber-planner! It could be an egg, a piece of tagboard or an earthworm. All required materials are listed on pages 9-10 for easy reference. Be prepared. A few items need to be ordered as much as 3 weeks in advance.
5. Read the "For My Notebook" section to yourself once so you know how to pronounce the new words in it. Pronunciation of all words that I think you might not know are included right there so you can read them as you go, but you may need to look up something I haven't thought of. I'm sorry I don't know you that well.
6. Curl up under that weeping willow or in front of the fire and read the "For My Notebook" page to your child, even if they can read by themselves. There is one for each major topic. Pause to do whatever it tells you to—looking at a fish, finding a spider, etc. The notebook pages are written to your child and should be saved in his own notebook. They are purposely short so you can go right from the notebook page to the first activity if you would like.
7. In a couple days do the second activity, or do the story page on day one and each activity on a separate day for 3 shorter doses of science each week.
8. Follow up with a day of nature journaling, reading from the extra reading list or drawing. Drawing is an important skill for many scientists. Have fun. And did I mention you should have fun?

SOME UNIQUE HELPERS IN THIS BOOK

ON THE STORY PAGE:

1. All student pages have a boxed outline around the material presented. That way it is easy to identify what is for the child and what is for the teacher. 2. You'll find all new vocabulary words underlined
3. All vocabulary that has appeared previously will be in italics. These will also have a reference at the bottom of the page to tell you which notebook page to go to in order to read about them again. You'll notice most of the new vocabulary doesn't have a classic dictionary definition. Instead the explanation is usually given in context so it is "felt" and not memorized.
4. All vocabulary is also listed in alphabetical order at the back of the book. This way, if you want a "dictionary definition" you can find it there.

ON THE INSTRUCTIONS PAGE:

1. The instruction page is really for the parent/teacher, but the procedure is written as if for the student. I have no way of knowing how capable your child is, so I leave it up to you to decide which steps you want your child to participate in (disclaimer).
2. Each instruction page includes a prompt to read aloud to your child. This is for the parent who wants it. If you hate those prompts, don't sweat it. I'll never know if you read them to your child or not.

WHAT'S THE BIG IDEA?

Whenever you study a subject you have main ideas to learn and little details that are a nice bonus. It's true that science has a lot of new vocabulary and information. If you are using a classical education approach to teaching, you will cover every subject three times throughout your child's education. Because of this, don't sweat the small stuff. There are many challenging words in here that are used because they are the right words and after hearing them over and over they will "sink in." They are not here for your child to memorize the first time around.

This outline gives you the big idea that your child should get from each section and the small stuff that is an added bonus. If you and your child are timid scientists, just have fun as you try to learn the big ideas. If you and your child have a strong science background, work on learning the small stuff as well as the big ideas. Use these difficult words and science concepts gently, not with force and your child will enjoy his science experience.

BI = BIG IDEA SS = SMALL STUFF

WHAT IS LIFE:

BI = Living things are made of cells
You can tell living things from nonliving things by what they can do

SS = Living things all 1. Take in energy 2. Get rid of waste 3. Move 4. Grow 5. Reproduce 6. Have circulation 7. Have respiration 8. Respond to their environment 9. Are made of cells

LIVING THINGS ARE MADE OF CELLS:

BI = Living things are made of cells.
Cells come in many different sizes and shapes but are usually too small to see.

SS = Plant cells are usually rectangular and animals cells are typically rounded.
Plant cells have chloroplast that make the plant green and make food for the plant.
The nucleus is the control center of the cell.

SKELETAL AND MUSCULAR SYSTEMS:

BI = Bones support you and make up the skeletal system.
Muscles move you and make up the muscular system.
Muscles can only pull, not push. Therefore they work in pairs to pull back and forth.

SS = Some of the major bones are the skull, ribs, humerus, femur, vertebrae, pelvis and patella.

CIRCULATORY SYSTEM:

BI = Your circulatory system is made of your heart, arteries and veins.
Your heart is a muscle that forces blood to every part of your body.
Blood carries food, water and oxygen to the cells.
The harder you exercise, the harder your heart works.
Blood is a mixture of 4 different things.

SS = Arteries carry blood away from your heart. Veins carry blood back to your heart.
White blood cells fight bacteria. Red blood cells carry oxygen. Plasma carries food.
Platelets seal up cuts to stop the bleeding.

RESPIRATORY SYSTEM:

BI = You breathe oxygen into your lungs.
This oxygen goes into your blood and then to every part of your body.
The harder you exercise the more oxygen you need so the faster you breathe.

SS = Air goes into your trachea and then to your lungs.
Carbon dioxide is traded for the oxygen in your lungs. You breathe out carbon dioxide.

DIGESTIVE SYSTEM:

BI = Your teeth, esophagus, stomach and intestines make up your digestive system.
Your digestive system is much longer than you are tall. It is twisted and coiled to fit into your body.
Your digestive system takes the food you eat, saves the nutrients and water that your body needs and sends the rest back out of your body.

SS = The digestive system uses special chemicals to help dissolve food.
Your body needs a variety of nutrients so eating a balanced diet is important for staying healthy.

NERVOUS SYSTEM:

BI = Your brain, spinal cord and nerves make up your nervous system.
Your nervous system's job is to take in information and to tell your body what to do about the information.
Your five senses gather information. With your five senses you see, hear, smell, taste and feel the world around you.

SS = People rely on some senses more than others. All animals have different senses that they favor.

GROWTH AND GENETICS:

BI = Your body grows in stages.
Special messages in your cells determine how you will look and grow.
You inherit characteristics from your parents.
Except for identical twins, no two people are exactly alike.

SS = Genes carry the messages you inherit from your parents.
You get a mix of genes from your mom and dad so you may show characteristics from each.
Everybody has a different set of fingerprints. No two people's are alike.

SIX KINGDOMS:

BI = Living things are grouped based on their characteristics.
Grouping living things helps us name, understand and communicate about them.

SS = There are six kingdoms or groups of living things. Two are the plant and animal kingdoms.
Every living thing (discovered) has a scientific name which tells us its species.

CNIDARIA:

BI = Some simple animals have no organs and look like plants.

SS = Cnidarians are simple animals much like a hollow sack with stinging tentacles.
Corals, sea jellies and sea anemones are cnidarians.
Sea jellies go through metamorphosis and change from polyp form to medusa form.

WORMS:

BI = Worms are long like a string and have no legs.
Earthworms are very helpful to people.

SS = Worms can be flat, round or segmented.
Worms have organs inside.
Earthworms can tell many things about their environments.

MOLLUSKS:

BI = Soft bodied animals like snails and octopi are grouped together.
Some of these soft bodied animals have shells.

SS = Soft bodied animals are called mollusks. Most have a shell either inside or outside.
Mollusks include snails, slugs, octopi, squids, clams and oysters.
Mollusks have complex organs like people do.

ECHINODERM:

BI = Sand dollars, sea stars and sea urchins have spiny skin and are grouped together.
Spiny skinned animals have bodies that can be divided into 5 parts

SS = Spiny skinned animals are called echinoderms.

ARTHROPOD:

BI = Insects, spiders and lobsters all have jointed legs and divided bodies.
Insects have 6 legs and 3 body parts and most insects have wings. Spiders have 8 legs, 2 body parts and no wings. Lobsters and isopods have more than 8 legs.
Insects and other animals go through metamorphosis.

SS = Arthropods all have jointed legs, segmented bodies and an exoskeleton.
The major groups of arthropods are insects, arachnids and crustaceans.
Stages of complete metamorphosis are: egg, larvae, pupae, adult.
Spiders, ticks and scorpions are all arachnids.
Crustaceans have gills and most live in water.

CHORDATA:

BI = Invertebrates have no backbones. Vertebrates have backbones.
Animals with backbones are fish, amphibians, reptiles, birds and mammals.

SS = Backbones are called vertebrae. Vertebrae protect your spinal cord.

FISH:

BI = Fish have fins and scales, lay eggs and breathe through gills.
Fish are cold blooded.

SS = Cold blooded animals have the same temperature as the surrounding air or water.
Fish float at different depths by adjusting the amount of air in their swim bladders.

AMPHIBIANS:

BI = Frogs, toads and salamanders are amphibians. Frogs and toads go through incomplete metamorphosis.
Amphibians are cold blooded and lay eggs. When they are young they have fins and gills but when they are grown they have legs and lungs.
Amphibians must stay near water.

SS = Amphibians must stay near water because they will dry out. Their skin is porous because they breathe through it.
Amphibian eggs must be laid in water because they have no shell.

REPTILES:

BI = Lizards, snakes, turtles and alligators are reptiles.
Reptiles have scales and lungs, are cold blooded and most lay eggs.

SS = Reptile eggs have a leathery shell so they can be laid away from water.
Reptiles can live far from water because their skin isn't porous.

BIRDS:

BI = Only birds and mammals are warm blooded.
Birds have beaks, feathers and wings and lay eggs with hard shells.
Birds' feet, beaks and body shapes tell a lot about how the bird lives and where.

SS = Feathers are for flying and also help the bird stay the proper temperature. Birds keep their feathers waterproof by spreading special oils on them.

MAMMALS:

BI = Mammals nurse their young, don't lay eggs (except for 3 exceptions), have hair or fur and are warm blooded.
People are mammals.
Mammals have ways to stay warm in winter and cool in summer.
Good common names of animals help us to identify them.

SS = "Mamma" is latin for "breast." Mammals have breasts for nursing their young. Polar bears, whales and other marine mammals have thick blubber to keep themselves warm.

PLANT KINGDOM:

FLOWERS:

BI = Flowers attract animals to pollinate them.
Seeds are made when flowers are pollinated.

SS = The female part of the flower is the pistil. The male part is the stamen. The petals attract pollinators and the sepals protect the flower before it opens.
Flower parts (petals, stamens etc.) vary greatly in number and size.

SEEDS:

BI = Seeds grow into plants.
Some seeds split into two parts, some stay as one.
The seed is a tiny baby plant and a lot of stored food to start it growing.
Seeds have interesting ways to spread around so they find space and water to grow.

SS = Monocots have one seed part, dicots have two. Each seed part is a cotyledon.
Each seed has a seed coat, an embryo and one or two cotyledons.
Seeds must soak up water before they can sprout.
Seeds spread by wind, water, sticking and being eaten as part of a fruit or berry, and then get dropped in a new place.

LEAVES:

BI = Leaves contain chloroplasts which make the plant green.
Chloroplast take sunlight and make energy.
Plants that don't get light cannot grow very well.
Leaves look solid green but are really just dots of green chloroplasts.

SS = Without plants there would be no life. Animals need plants in order to survive.
Without sun, chloroplasts turn yellow.

STEMS AND ROOTS:

BI = Roots suck water up from the ground and hold the plant up.
Stems have tubes that carry food and water through the plant.

SS = Water can only enter a plant through the roots.

MATERIAL LIST

Material is listed by unit in the order in which it is first needed. + means it will be needed for later units also. The amounts listed are totals for the entire course. Most items are common household items. * means it requires some explanation. Ordering hints or explanations are given on page 10.

REUSABLE EQUIPMENT / MATERIAL

UNIT			AMOUNT
1+		Plant with many healthy leaves	1
1		Rock to observe (optional)	1
1		Wild area to study (field, forest etc.)	
1		String	4 meters
1+		Meter/yard stick combination	1
1+		Clipboard	1
1+		Large clear jar with lid (32 oz. or more)	1 - 4
1+		Hand lens (Magnifying glass)	1
1	*	Field guides to local insects, animal tracks plants (optional)	
2+		Small plates	3
2+		Small round bowl (to hold 2 cups)	2
2		Small square pan (to hold 2 cups)	1
3+	*	Glue gel (like Elmer's School glue gel)	1
3+		Hole punch	1
3+		Scissors	1
3+		Colored pencils, crayons, markers	
3+		Stapler	
4		White or gold pen (to mark on black)	1
4+		Ruler with centimeters and inches	1
5		Tennis ball	1
5+		Watch with second hand	1
5+		Spoons (at least one metal)	2
6		Long tube (wrapping paper, vacuum cleaner...)	1
6		Small colored counters- blue and red	10 each
7+		3" x 5" index cards (white)	7 or more
8+		Small jars with lids (see-through)	3
8		Dry erase marker	1
11+		Various art supplies- stamps, stickers etc.	
13+		Flashlight	1
13		Red cellophane (optional)	approx. 4" x 4"
13		Pitchfork (to dig up worms)	1
13+		Mister with water	1
13+		Tray with sides (9" x 9" pan, food storage container etc.)	1
13		Sandpaper- rough	approx. 9" x 9"
14+	*	Gram scale or triple beam balance	1
15+		Kitchen knife	1
16+	*	Butterfly house (bought or made)	1
18		Small piece of rotting log for critter keeper	
18		Chalkboard or whiteboard (or butcher paper)	1
18		Heating pad or hot water bottle	1
18+		Zipper type plastic bag- (1-1 gallon) (2 small)	
20		Fish net - small for netting goldfish	1
20+	*	Thermometer- science type	2
20		Thermometer - sterile, human type	1
20+	*	Glass eyedropper	1
21		Metal Slinky toy	1
21		Metal pie pan	1
22		Cookie sheet	1
22		Waterproof marker	1
23+		Zoo or wildlife center to study	
23	*	Binoculars (optional)	1
24		Samples of animals studied: shells, rubber or plastic critters, live insects or worms in vials, pictures, dried coral or sea star...	
25		Tweezers (optional)	1
26		Computer with internet access (optional)	
26		Sunday cartoon section of the newspaper	1
28		Identical, upright plants in small pots	2
28		Clear drinking glasses	3
28		Clock	1

PERISHABLES

UNIT		AMOUNT
2	Chicken egg	1
2	Light colored gelatin (lemon)	6 oz. box
2	Green grapes	20
2	Strawberries or orange slices	2
3+	12" x 12" colored cardstock	8
3	12" x 18" construction or other heavy paper	1
3+	Yarn- any color	about 12 yards
4	8 1/2" x 11" black construction paper	1
4	8 1/2" x 11" tagboard or thick paper	1
4+	Metal fastener (brad)	2
4	Thin string or thick thread	28" long
4+	Tape	
5	Light corn syrup	1/2 cup
5	Red Hot candies	1/2 cup
5+	Dry large lima beans	6
5	Dry lentils or split yellow peas	1 Tbl
6+	8 1/2" x 11" white paper	6
8	Paper towels	15
8	Items to test senses- may include: coins cotton swab, paper clips, vinegar, garlic, chocolate chips, banana, orange, bells	
8+	Paper sacks (4 lunch size) (1 large)	4
12	Styrofoam cups (stacking)	4
13+	Sand	about 2 cups
13+	Soil	about 2 cups
13	Oatmeal (dry)	about 1 cup
13+	Aluminum foil	about 14"
13	Vinegar	1/4 tsp.
14	Snail food (lettuce, cabbage, ivy, cucumber)	
15	Banana (unpeeled)	1
18	Ice	
19	Card stock or construction paper 18" x 12"	1
19	Magazines with animal pictures (vertebrate and invertebrate) - cuttable	1 - 4

UNIT	PERISHABLES (CONT.)	AMOUNT
21	Salt	1 Tbl.
21	Manila file folders	2
23	Vegetable oil	2 Tbl.
24	Shortening or lard	1/2 cup
24	Butcher paper	approx. 3' x 6'
25	Flower (gladiolus is best)	1
26	Corn seeds	11
26	Dried kidney beans	11
26	Socks- big, old	2
26	Cranberry and other fruit (to show seeds)	26
	Seed selection- maple, dandelion, elm	
28	Stalk of celery with leaves on top	1
28	White carnation	1
28	Food coloring- red and blue	
28	Salad dressing	
28	Salad fixings of your choice	

CRITTERS

UNIT		AMOUNT
13+	*Earthworms (dug up or bought)	6
14+	*Garden snails	5
16+	* Butterfly larvae (gathered or bought)	5
17+	* Spider to observe (real or photos)	1
18+	* Isopods (roly-polies, pill bugs)	10
20	Fish - goldfish works well	1

Please Note: All experiments are designed to teach about the animals without harming them.

SPECIFIC EQUIPMENT: HINTS AND ORDERING

1. **Field guides to insects, plants, birds etc.:** A list of good field guides is provided at the bottom of the list of reading suggestions on page 15.
2. **Glue gel:** We call this "blue glue" because it is blue. It is the "Elmer's washable school glue gel." It sticks better than a glue stick and isn't as messy as white glue. It is available at almost any office or school supply store.
3. **Gram scale or triple beam balance:** Either of these will weigh items in grams. Gram scales are easier to use. Triple beam balances are frequently used in many laboratories so it may be worth purchasing one and learning how to use it. Gram scales are available at many drug stores. Either can be purchased from a science supply store.
4. **Butterfly house:** can be made from a box with netting over one side. Look for specific instructions on the internet. Plan about 3 weeks ahead if you want to buy one. You can purchase one from Insect Lore, and it will come with a certificate to get 5 butterfly larvae. You just send in the certificate (and a few dollars for shipping) about a week before you are ready to start your butterfly unit.
5. **Thermometer:** A good science thermometer goes down on its own like a room thermometer. Remember a human thermometer must be shaken down. You will need a thermometer that isn't a permanent part of a decoration and that has the temperatures marked right on it. Some food thermometers are built right but don't go very low as they are used for hot cooking. You need one to go down to 30° F or 0°C. If all else fails, order one from Nasco or another science supply store.
6. **Glass eyedropper:** This should be easy but it isn't always. Drug stores usually offer big, bulky plastic ones but these won't do what you need. Find a nice, small, old-style, glass thermometer. I found one at an old-style pharmacy. Otherwise, order one from a science catalogue.
7. **Binoculars:** 7 X 35 is a nice choice to get for birdwatching. I would avoid the $10 ones as you won't see much but you don't have to spend $100 on a set for your kids either. Whatever you get, look through them first. The optics can vary from one pair to another. Look in the sports or electronics department .
8. **Garden snails:** There is a care sheets provided (pg. 11) that gives hints on finding and housing snails.
9. **Earthworms, butterfly larvae, spiders (optional) and isopods** (pill bugs): Finding and housing them is explained in the labs where they are first needed.

SOURCES FOR SCIENCE MATERIALS

Home Science Tools (vast array of science materials from frogs to magnets)

Offers a materials kit for RSO Life: www.hometrainingtools.com/r-e-a-l-science-odyssey/c/2055/

Insect Lore (fun equipment for learning about and raising critters – crabs to ants)

P.O. Box 1535, Shafter, CA 93263 Phone: 1-800-LIVE BUG www.insectlore.com

COLLECTING AND HOUSING GARDEN SNAILS

FINDING: Garden snails are easiest to find in lush gardens, woodlands or parks right after a rain (or watering). They also are easiest to find at night. If you don't have a garden, or know anybody with a snail problem (lucky you!) try a park or nursery where they grow plants.

HOUSING: Snails can be kept in a solid walled critter keeper, large jar or small aquarium. Snails climb well so make sure the lid is on and make sure it has plenty of breathing holes. Put a little soil in the bottom, and add a nice piece of rotting log. Keep the soil moist, but not wet by misting daily. Add a small dish of water to keep the air humid.

FEEDING: If you can, add leaves that you found the snail on. Also feed lettuce and cucumber. Avoid any leaves that have been sprayed with chemicals.

ADDITIONAL READING SUGGESTIONS

highly recommended * more fun than science

Some great, grammar stage books you could use for several sections of this book are given below with abbreviations. They are listed under each topics with the appropriate page numbers for that topic.

Unit 2: THE CELL

#Greg's Microscope - Millicent E. Selsam

Unit 3: THE HUMAN BODY

#What's Inside You - Susan Meredith (Usborne Starting Point Science)
#Science and Your Body - Rebecca Heddle (Usborne Science Activities)
SOMEBODY: Five Human Anatomy Games - Fun games to help teach anatomy
The Magic School Bus Inside the Human Body - Joanna Cole
Me and My Amazing Body - Joan Sweeney
Your Insides - Joanna Cole

Unit 4: THE SKELETAL AND MUSCULAR SYSTEMS

The Skeleton Inside You - Philip Balestrino

Unit 5: CIRCULATORY SYSTEM

Your Blood and Its Cargo - Sigmund Kalina
A Drop of Blood - Paul Showers

Unit 7: DIGESTIVE SYSTEM / NUTRITION

Good Enough To Eat - Lizzy Rockwell
Eat Healthy Feel Great - William Sears, M.D, et. al.

Unit 8: NERVOUS SYSTEM

The Magic School Bus Explore the Senses - Joanna Cole
My Five Senses - Aliki

Unit 9: GROWTH AND GENETICS

When an Animal Grows - Millicent Selsam (about non-human animal growth)
Me and My Family Tree - Paul Showers

Unit 10: CLASSIFYING LIFE

Usborne Illustrated Encyclopedia of the Natural World
QUICKPIX Not Just An Animal Game - Aristoplay
#Benny's Animals and How He Put Them in Order - Millicent E. Selsam
Animals of the Sea - Millicent Selsam

Unit 12: CNIDARIA

Usborne Illustrated Encyclopedia of the Natural World Jellyfish - Leighton Taylor

Unit 13: WORMS

Usborne Illustrated Encyclopedia of the Natural World Worm Day - Harriet Ziefert
Earthworms - Dorothy Childs Hogner

Unit 14: MOLLUSCA

Usborne Illustrated Encyclopedia of the Natural World

#*Are You a Snail*- Judy Allen

Unit 15: ECHINODERMATA

Usborne Illustrated Encyclopedia of the Natural World

ARTHROPODA:

Unit 16: Insects:

Usborne Illustrated Encyclopedia of the Natural World

Usborne Complete First Book of Nature

The Magic School Bus Gets Ants in Its Pants - Joanna Cole

The Magic School Bus Inside a Beehive - Joanna Cole

Zoobooks: Butterflies - The Knowledge Company (distributor)

Butterflies - Karen Shapiro (Scholastic "Hello Reader!" book)

The Butterfly House - Eve Bunting

#*Waiting for Wings* - Lois Ehlert

Are You a Butterfly? - Judy Allen

Are You a Ladybug? - Judy Allen

Are You an Ant? - Judy Allen

**Crickwing* - Janell Cannon

Ma Jiang and the Orange Ants - Barbara Ann Porter

Terry and the Caterpillars - Millicent Selsam

Unit 17: Arachnids:

Usborne Illustrated Encyclopedia of the Natural World

Are You a Spider? - Judy Allen

**Charlotte's Web* - E. B. White

Unit 18: Crustaceans:

Usborne Illustrated Encyclopedia of the Natural World

A House for Hermit Crab - Eric Carle

Pagoo - Holling C. Holling

**Kermit The Hermit* - Bill Peet

CHORDATA

Unit 20: Fish:

Usborne Illustrated Encyclopedia of the Natural World

Usborne Complete First Book of Nature

#*Plenty of Fish* - Millicent Selsam

**Swimmy* - Leo Lionni

Unit 21: Amphibians:

Usborne Illustrated Encyclopedia of the Natural World

#*Song of La Selva* - Joan Banks

**Frog and Toad Are Friends* - Arnold Lobel

**Frog and Toad Together* - Arnold Lobel

*Frog and Toad - Arnold Lobel

Unit 22 Reptiles:

Usborne Illustrated Encyclopedia of the Natural World

Imagine You Are a Crocodile - Karen Wallace

The Moon of the Alligators - Jean Craighead George

**Verdi* - Janell Cannon

Unit 23: Birds:

Usborne Illustrated Encyclopedia of the Natural World
Usborne Complete First Book of Nature
#Tony's Birds - Millicent Selsam
See How They Grow - Owl - Kim Taylor
**Make Way for Ducklings* - Robert McCloskey

Unit 24: Mammals:

Usborne Illustrated Encyclopedia of the Natural
World Usborne Complete First Book of Nature
#Nature Detective - Millicent Selsam
**Stellaluna* - Janell Cannon

THE PLANT KINGDOM

Usborne Illustrated Encyclopedia of the Natural World
Usborne Complete First Book of Nature
**Franklin Plants a Tree* - Paulette Bourgeois

Unit 25: FLOWERS

Usborne Illustrated Encyclopedia of the Natural World
Usborne Complete First Book of Nature
The Reason For a Flower - Ruth Heller

Unit 26: SEEDS

Usborne Illustrated Encyclopedia of the Natural World
#Seeds and More Seeds - Millicent Selsam
The Tiny Seed - Eric Carle
**The Carrot Seed* - Ruth Kraus

Unit 27: LEAVES

Usborne Illustrated Encyclopedia of the Natural World

Unit 28: STEMS AND ROOTS

Usborne Illustrated Encyclopedia of the Natural World

OTHER LIFE SCIENCE TOPICS:

EVOLUTION / NATURAL SELECTION

Usborne Illustrated Encyclopedia of the Natural World
The Story of Life on Earth - Margaret Munro
Evolution - Joanna Cole
Evolution- The Beast in You (an activity book) - Marc McCutcheon

PREHISTORIC LIFE

Usborne Illustrated Encyclopedia of the Natural World
The Story of Life on Earth - Margaret Munro
Wild and Wooly Mammoths - Aliki *Dinosaur*
Bones - Aliki
Digging Up Dinosaurs - Aliki
My Visit To The Dinosaurs - Aliki
Patrick's Dinosaurs - Carol Carrick
What Happened To Patrick's Dinosaurs? - Carol Carrick

REPRODUCTION

The Human Body: A First Discovery Book Scholastic
Everybody Has a Bellybutton - Laurence Pringle
How You Were Born - Joanna Cole
Before You Were Born - Ann Douglas

FIELD GUIDE SUGGESTIONS:

INSECTS

Insects (Peterson Field Guide Series) - Donald J. Borror / Richard E. White

BIRDS

Field Guide to the Birds of North America (National Geographic Society) - Shirley L. Scott
Stokes Field Guide to Birds (by region) - Donald and Lillian Stokes
Field Guide to Western (or Eastern) Birds (Peterson Series) - Roger Tory Peterson
Birds of North America (Golden Guide) - Chandler S. Robbins et al

REPTILES AND AMPHIBIANS

A Field Guide to Western (or Eastern) Reptiles and Amphibians (Peterson Field Guide Series) - Robert C. Stebbins

MAMMALS

A Field Guide to the Mammals of America North and Mexico (Peterson Field Guide Series) - William H. Burt / Richard P. Grossenheider

CARE OF UNUSUAL "PETS"

#Pets in a Jar - Seymour Simon
Animal Care From Protozoa to Small Mammals - F. Barbara Orlans
Creepy Crawlies and the Scientific Method - Sally Kneidel

WEBSITE SUGGESTIONS

WHAT IS LIFE?

What is "living"? www.hitchams.suffolk.sch.uk/key/plantand.htm

THE HUMAN BODY

All body systems www.innerbody.com/htm/body.html

SKELETAL AND MUSCULAR SYSTEMS

Info on bones http://vilenski.org/science/humanbody/hb_html/skeleton.html

CIRCULATORY SYSTEM

http://sln.fi.edu/biosci/biosci.html

RESPIRATORY SYSTEM

Great info http://www.sk.lung.ca/content.cfm/kids http://yucky.kids.discovery.com/noflash/body/pg000138.html

DIGESTIVE SYSTEM

Fun nutrition site http://www.nutritionexplorations.org/kids/main.asp

NERVOUS SYSTEM

Senses- info, labs etc. http://faculty.washington.edu/chudler/chsense.
html Nervous system info http://faculty.washington.edu/chudler/neurok.html

GENETICS AND GROWTH

Explains genetics and DNA www.genecrc.org/site/ko/kola.htm

CNIDARIA

Click to see sea jellies www.mbayaq.org/efc/living_species/default.asp?hOri=1&group=2
Coral reef coloring www.surfnetkids.com/cgi-local/go.cgi?
http://www.reefrelief.org/kids/

Reef info, map etc. www.enchantedlearning.com/biomes/coralreef/coralreef.shtml

WORMS

Earthworm answers sps.k12.ar.us/massengale/earthworm%20facts.htm

Worm and composting info http://yucky.kids.discovery.com/noflash/worm/pg000102.html
Links for more worm info www.hitchams.suffolk.sch.uk/key/worms.htm MOLLUSCA
Without shells www.hitchams.suffolk.sch.uk/key/slugs,.htm
With shells www.hitchams.suffolk.sch.uk/key/snails&.htm

ARTHROPODA

Classification, projects members.aol.com/YESedu/welcome.html
Art project www.hhmi.org/coolscience/butterfly/index.html
Lesson plans, insect housing insected.arizona.edu/uli.htm

Facts, crafts, insect housing www.uky.edu/Agriculture/Entomology/ythfacts/entyouth.htm

Butterfly metam., info www.mesc.usgs.gov/resources/education/butterfly/bfly_start.asp

Spider info, poems, lessons (WARNING: graphic bite photos)

arthur.k12.il.us/arthurgs/spidlink.htm

Crustaceans www.hitchams.suffolk.sch.uk/key/shrimps&.htm

Click on the crabs to view and see information

www.mbayaq.org/efc/living_species/default.asp?hOri=1&group=2

ECHINODERMATA

Great echino info	www.enchantedlearning.com/subjects/invertebrates/echinoderm/

CHORDATA

Vertebrate puzzles	nationalzoo.si.edu/audiences/kids/

Fish:

Fish classification/activities	www.geocities.com/sseagraves/fishclassification.htm

Amphibians:

Frogs	allaboutfrogs.org/froglnd.shtml

Reptiles:

Sea turtle rehab., info	www.seaturtlehospital.org/index.htm

Birds:

Info, coloring, activities etc.	www.enchantedlearning.com/subjects/birds/
Activities, diversity, anatomy etc.	www.nhm.org/birds/home.html
Live webcams of nesting raptors (nesting season)	www.newyork.org/webcams/webcams.htm

Mammals:

Activities, diversity, anatomy etc.	www.nhm.org/mammals/home.html
Record breakers- fattest, smallest	www.abdn.ac.uk/mammal/records.htm
Whale info, activities etc.	www.zoomschool.com/subjects/whales/
Links to whale lessons/activities	members.aol.com/donnandlee/Features.html#Whales

THE PLANT KINGDOM

Summary of the plant parts	www.hhmi.org/coolscience/vegquiz/plantparts.html
Plant parts mystery	www.urbanext.uiuc.edu/gpe/index.html

SEEDS

More seed activities (parent pg)	www.thirteen.org/edonline/nttidb/lessons/cb/plantcb.html
Great seed picture	www.theseedsite.co.uk/seedparts.html

KEEP A SCIENCE JOURNAL

Nature is beautiful and amazing, but it is not a video game. The action in nature is far more subtle (usually) but well worth looking and waiting for. Life, death, greed, compassion and romance are all just waiting to be discovered by the observant and patient. Writing it all down is a skill worth developing. Sketching and saving samples of the common things you find along the way all add up to give you a journal you can be especially proud of and will want to cherish your whole life. For a satisfying nature experience, grab that journal and that pencil, maybe a hand lens (magnifying glass) and a pair of binoculars and hit the trail, or the backyard or park. You don't need to have a plan when you go out; but to get you started, we have provided, in no particular order, a few hints as to what might go into a nature journal. These are just ideas to get your creative juices flowing. Writing in your journal should be like writing to yourself. Don't worry about spelling, sentence structure or grammar. Now go out, observe, draw, listen, describe, compare—journal! Remember, the more you put into it, the more you will get out of it.

For each journal entry remember to put:

• The date • The weather (temperature, clouds, rain, wind, etc.) • Who you went with

AUTUMN:

- Make a leaf rubbing. Place a leaf under one journal page. Choose a color that matches the leaf. Rub the crayon over the paper where it covers the leaf. In your journal, describe the plant it came from and what it looks like this time of year. Repeat this in the spring.
- Sit by a tree. What things might that tree be used for if it were cut down? What might it be used for if it were to grow? In its lifetime, how many more animals might be able to enjoy it as it grows? What species is the tree? If you know, write it down. Sketch a picture of the tree.
- Use your sense of hearing. Close your eyes. What do you hear? What do you think it means? Write a factual or fictional story about what you hear.
- Find an insect of any kind. Draw it. Describe what it is doing. Don't disturb it in any way. Watch what it does when it is not disturbed.
- Check out a clear night sky. Can you find any constellation? Satellites? Bats?
- Pick up two different rocks. Describe them. Compare weight, shape, texture, color...
- This is a safe time to find a nest as they will be empty by now. How is it made? What material was used? How was it put together? What feathers are nearby?
- Describe the weather changes you see. Draw the clouds, take the temperature, feel the wind.
- Which trees are losing their leaves first? Which in clumps? Does size matter? Species (kind of tree)? Location?
- Draw a tree in the autumn. Remember which one it is. Draw the same tree in the winter, spring and summer.

WINTER:

- Describe a deciduous tree (one that loses its leaves in the winter) and an evergreen tree (one that stays green all a year). How are they shaped? How different do they look this time of year? Is one easier to see in the winter?
- Look for predators (meat eaters). Winter is a hard time of year for them. Look for hawks, owls, weasels and coyotes. How do they look? What do you see that they could be finding to eat? - There is a different kind of silence when it is snowy or foggy. Sit, listen and describe a snowy or foggy day. What do you hear? How does it sound different than it does on a clear summer day?
- Finish the sentence "I'm so cold I could..." How do animals and plants survive the cold?
- Describe and sketch the nearest mountains. What covers them? Who uses them?
- When it rains, watch how the water washes dirt down a hill. Can you see it start to form tiny rivers? What does the water do when it meets up with sticks, rocks and leaves?
- Use your nose after a rain. How does the world smell? How is that different than usual?
- Put a stick in the ground. Measure its shadow morning, noon and evening. Draw your stick clock in your journal. Show where the shadow is at different times of the day. - How do birds look in winter? Describe what they do, where they perch, what they say.

SPRING:

- Find a pond or lake. Sketch it and the surrounding area in your journal. Color in where you would expect to find fish in blue, amphibians in green and reptiles in brown.
- Find a pond or puddle. Look closely for animal tracks. Sketch them in your journal. Can you determine where each animal went and what it was doing? Tell a story of the events that lead to the footprints.
- Find a flower, the bigger the better. Watch and see if any animal visitors come to pollinate it. Write and sketch about what you see.
- Carefully draw a flower. How is it built to draw in pollinators? Guess by its size and shape what animals might pollinate it.
- Listen to the birds. Birds sing to claim a territory and advertise for a mate. Describe their songs. Are they sweet? Flat? Enthusiastic? Impressive?
- Try to write a bird's song in words. A quail's call is written as "Chi-**ca**-go." A chickadee's name comes from its call—"Chicka-dee-dee."
- Don't forget to journal at night. Describe the life attracted by a porch light at night.
- Use your nose. What can you smell? Sit and smell, then look for things that make smells—flowers, trees...
- Find a butterfly. Describe its color, patterns, flight. Where are its wings when it is resting? Now find a moth (easiest at night) and compare.

SUMMER:

- Go outside. Watch seeds move, fly, pop open. Write and sketch about what you see.
- Find a column of ants. Don't bother them in any way. Where are they going? What are they carrying? Can you find individual ants with different jobs? How quickly are they moving? Follow and time one ant to find the speed of the column.
- Sit quietly and observe the birds. What birds do you have in your area? Draw, describe and name them when you can.
- Walk through a field of stickers. Pull them out of your socks, sketch and describe them. How many ways can you find that seeds have of sticking? Find the seed in each sticker. - Finish the thought "I'm so hot I could..." How do animals and plants adjust to the heat?
- Describe the plants and animals near a river or stream. How is life different here than just a little ways away from the water?
- Find an animal path in nature. What animals do you think use it? How worn is it? How tall are the animals that use it? Look for prints, hair, droppings, feathers.
- Find a feather. Draw it and describe its shape, structure, texture, color. Flap the air with it. Is it noisy or silent? Guess what bird might have left it there.
- Visit a flower in the morning, at noon, in the evening and again at night. How does it change throughout the day? Notice its texture, droop, color, how open it is.

NAME ______________________ DATE ______________

For my notebook

What Is Life?

You are alive, aren't you? How about dogs and cats? They are alive too. You probably just know this without knowing exactly why. How about worms, trees or coral? You are going to be learning about a lot of different living things but first let's learn to tell if something is alive. Scientists have found that all living things are made of tiny building blocks called cells (selz). Just like you might build a castle out of Legos of different sizes and shapes, living things are made of stacks of different cells. Unfortunately, most cells are too tiny to see so we will learn other ways to tell if something is a living thing. Think of your own body when you read this list of living characteristics.

All living things 1. take in energy (we do this by eating) and 2. get rid of the waste. They all 3. move, although some, like plants, move very slowly. All living things 4. grow and can 5. reproduce or make babies. All living things have some type of 6. circulation- like our blood moving in our bodies, and some type of 7. respiration- like our breathing. The last thing is, all living things 8. respond to what's around them. Many objects that aren't alive do some of these things but only living things do all of them. For instance, a car isn't alive but it moves and gets rid of waste (exhaust). Think about your own body now. Can it do all of the things on this list? Yes, it can because you are alive!

Living Lab #1: SIGNS OF LIFE - instructions

Material:

copy of lab sheet (1 page), pencil
plant and rock to observe (optional)

Aloud: All living things are made of building blocks called cells. Unfortunately, cells are usually too small to see without a microscope so we have to use other features like growth, respiration, a fancy word for breathing, and movement. In the activity today you will think about 4 things you have seen before and decide if they are alive by using a check list of characteristics. Living things will do all of the things on the list. Things that aren't alive can often do some of the things on the list. Some things, like circulation, are hard to see. Do your best and through the year as we study living things you will learn how to look for circulation, respiration and other signs of life.

Procedure:

Lab Day:

1. Before going through the check list of living characteristics, guess which of the four items you think are alive.
2. Do your best to fill in the chart.

HINTS FOR PARENTS:

FOR BIKE AND ROCK:
Let kids use their imaginations but guide them to understand bikes and rocks only move with help from gravity, being pushed etc., and they don't reproduce (make another whole bike or rock). They don't show circulation, respiration or a response to the environment but kids may come up with creative ways to say they do, which is fine.

PLANTS:
Take in energy = although they don't eat, plants get energy from the sun
Give off waste = plants give off oxygen (a waste product) just like we give off carbon dioxide.
Movement = plants move slowly to face the sun, some flowers and leaves open and close at night
Circulation = plants have tubes that carry water up and food down and throughout the plant. We will be looking at these later in the year, but for now you can check out the tubes when you "string" your celery.
Respiration = plants breathe in carbon dioxide and breathe out oxygen (the reverse of what people do)
Respond to Environment = plants lose their leaves in the winter, give off electrical impulses when cut, sprout when the rain and temperature are just right, etc.

3. Finish lab sheet.

Possible Answers:

1. The easiest characteristics are usually growth (it's easy to see your bike and a rock aren't growing) and reproduction. Even though a piece of rock splits off of a bigger rock, that piece will never grow up to make full sized rocks of its own.
2. If you look at a living thing under a microscope you will see it is made of cells.

Conclusion / Discussion:

- Obviously, people and plants are alive, but not bikes and rocks. Your chart might show a different answer, which is fine, as some characteristics are hard to see. The object isn't to get it right, as much as to start to look for living characteristics in other things such as sponges, corals, worms etc. This takes practice.

For More Lab Fun:

1. If this was easy and fun for the kids, challenge them to explain why a computer, a car, a river or a fire are not alive. All of these show living characteristics.

Living Lab [illegible]: SIGNS OF LIFE - Instructions

NAME ______________________________ DATE ___________________

Living Lab #1: SIGNS OF LIFE

Are the items on the line below alive? On the line next to each one write Y for yes or N for no.

MY GUESS BEFORE DOING THE LAB

People _____ Bike _____ Rock _____ Plant_____

On the chart below, put an X in the box if the object has the characteristic listed. If not, leave it blank. Put a ? if you're not sure.

CHARACTERISTIC	PEOPLE	BIKE	ROCK	PLANT
Takes in energy ex: (food, minerals)				
Gets rid of waste ex: (sweating, urinating)				
Moves (by itself)				
Grows				
Reproduces (can make babies)				
Circulation (stuff moves inside- blood, water etc.)				
Respiration (takes in and gives off gasses) Breathing				
Responds to environment ex: (moves towards the light, runs from danger)				

For something to be alive it must be able to do all of the things on the list. After doing the lab, do you think the things on the list are alive? On the line next to each one write Y for yes and N for no.

AFTER LAB

People _____ Bike _____ Rock_____ Plant_____

1. Some characteristics are hard to see- like circulation and respiration. Which characteristics from the list are the easiest to use?

____

2. If you could look at any living thing under a microscope, what would you see it was made up of? ______________________________

Living Lab #2: PLOT STUDY - instructions

Material:

copy of lab sheet (1 page), pencil
field, overgrown lot or other wild area at least 1 meter square to do your study
string: 4 meters long, marked at one meter intervals meter stick
clipboard and pencil
small clear jar or critter keeper
hand lens / magnifying glass (optional but recommended)
field guides of local insects, animals tracks, plants (optional)

Aloud: Today we will do a plot study. This means we will mark off an area, called a plot, outside and look closely at the living and nonliving things that make up that plot. Scientists do this to compare the features of one area with another. By doing this they can learn what kinds of areas certain plants and animals can grow in. It also helps to tell them when an area is becoming unhealthy. We will be looking for living and nonliving things in our study plot. You may have to look closely, but I'm sure that you will be surprised at how much life you will find! There are many living things so tiny we walk right over, and ON them without even knowing they are there.

Procedure:

Lab Day:

1. Choose a good, preferably wild area to study. A field, overgrown garden or empty lot works well. Try to avoid a manicured lawn. There won't be much plant diversity there.
2. Lay your string out in a square in the area you want to do your study. All study takes place in the square. No fair writing down what you see out of the square, even if it is neat-o, but your plot can go as high into the air or as deep into the soil as you wish.
3. Make a list of living and nonliving things you see in the plot. Every species should be listed on a separate line. Don't just put "weed." Put "Dandelion" or "small plant with yellow flowers and spiky leaves." Describe what you can't name. Use as much detail as possible. Under living, put the insects, worms, plants, leaves etc. For nonliving, list the different rocks you see, soil types, any man-made objects, water. Twigs and leaves were once living so they go there. Nonliving is not the same as dead.
4. After looking closely with a hand lens (you might be amazed at how pretty "dirt" can be), inspect under rocks, leaves and branches. These are great places to find beetles, worms and other interesting creatures. Every item turned over should be placed back exactly how it was so its inhabitants will not be displaced. Look up into the sky. Is there a tree above your plot? Birds flying over? The more you can get your children to just spend time looking, the more they will get out of this activity.
5. Choose a living and nonliving thing to draw. Put the living thing in the container while you draw it.
6. Finish your lab paper. For question 3, if you have a plot with a tree, estimate its height, otherwise measure.

Conclusion / Discussion:

- Discuss your list of living and nonliving things. What did you have more of? How would that be different on a beach or in the rainforest?
- Discuss the signs of life passing through. Did you find footprints, droppings etc.? Did you find any signs of human damage?

For More Lab Fun:

1. Start a nature journal. Write a description of the living and nonliving things you come across. Tape in samples of leaves or common plants you come across. Include drawings to help you remember what you have seen.

Living Lab #2: PLOT STUDY - Instructions

[illegible]

NAME ______________________________ DATE ____________________

Living Lab #2: PLOT STUDY

List the major living and non-living things in your study area.

LIVING	NON-LIVING

Signs of living things visiting my plot (droppings, tracks, partly eaten leaves...):

____

____

Signs of humans visiting my plot (footprints, writing, trash...):

____

____

Choose one living and one non-living thing from your plot to draw. Look very closely at them. Draw them in as much detail as possible.

LIVING	NON-LIVING

1. Choose one word to describe your plot (forest, grassy...)____________________
2. Choose one word for the soil in your plot (rocky, sandy...) ____________________
3. How tall is the tallest plant in your plot? ____________________

NAME ______________________________ DATE ____________________

For my notebook

LIVING THINGS ARE MADE OF CELLS

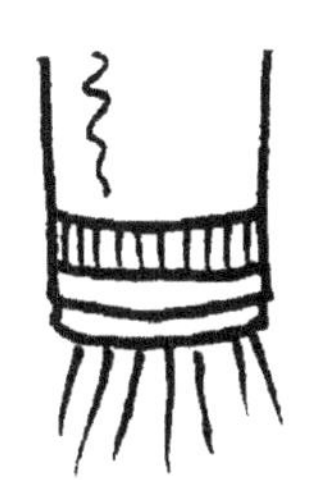

All living things are made of cells just like all snowmen are made of snowflakes. Most cells are too tiny to see just like snowflakes are but if you put enough together you can make something big enough to see. Some living things are made of only one cell but you are made of way too many to count. In your body there are many different kinds of cells to do lots of different jobs. Your blood has round, candy shaped cells that carry oxygen and blobby cells that can change shape to attack diseases in your body. Most animal cells are sort of round and most plant cells are rectangles that are held together like blocks in a wall to hold the plant up. Another difference between animal and plant cells is that plant cells have football shaped things called <u>chloroplasts</u> (**klor**-o-plasts). Have you ever heard that plants can make food from sunlight? The green chloroplasts in the plant cells do that job. There are so many of them that they make almost the whole plant look green. Most cells also have a <u>nucleus</u> (**new**-klee-us). This is a very important part of the cell. It is the control center. Just like your brain controls what your body does, the nucleus controls what the cell does, how it moves and how it looks.

Cell Lab #1: AN EGG IS A CELL - instructions

Material:

copy of lab sheet (1 page),
colored pencils: orange, yellow, blue, green and brown
chicken egg hand lens small dish

Aloud: What is every living thing made of? Cells, right? Most cells are far too tiny to see. It takes a lot of different types of cells to make all the different parts of all the different living things on earth. Some of those cells are visible without a microscope. We are going to look at one cell that is big enough to see without expensive equipment. An egg is a cell with a very special purpose- beginning a new life. Let's look and learn about cells and this special, life giving cell.

Procedure:

Lab Day:

1. Inspect the shell with a hand lens. It has very tiny holes to allow air and water to enter and leave the cell.
2. Gently break open the egg. Try to not break the yolk. Pour the contents and place the shell onto a small dish. Compare the egg to the egg diagram, finding the parts listed. Inspect the yolk very closely to find the very tiny, white blastodisc. It is a very important part of the cell and contains all of the cell parts except the food storage for the growing chick.
3. Use what you know and the clues from the chart to label the parts of the egg. Color the parts as indicated.
4. Draw the actual cell parts (hint: there are only 2) in the box provided.

Possible Answers:

#1 The yolk and the blastodisc are the actual cell parts.

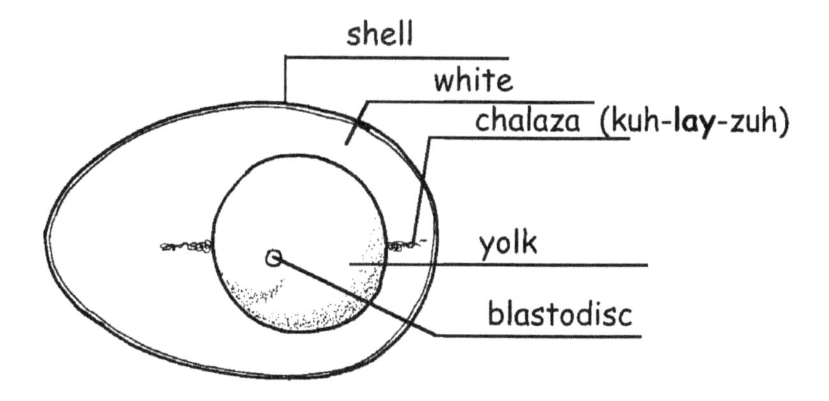

Conclusion / Discussion:

1. Why would this type of cell be so well protected? What might happen if it weren't surrounded by a shell and white gooey liquid?
2. Eventually a chick could hatch out of a fertilized egg. The yolk will have disappeared by then. Where do you think it might have gone? (To feed the growing chick)
3. How could you tell, just by looking that this is an animal cell and not a plant cell? (Plant cells are usually rectangular and have green chloroplasts)

For More Lab Fun:

1. Find out which animals lay eggs with a hard shell and which lay eggs without a shell.
2. Host a cell party. Have a cell (egg) and spoon race and a cell toss. Served deviled cells and cell salad sandwiches.

NAME ______________________ DATE ______________

Cell Lab #1: AN EGG IS A CELL

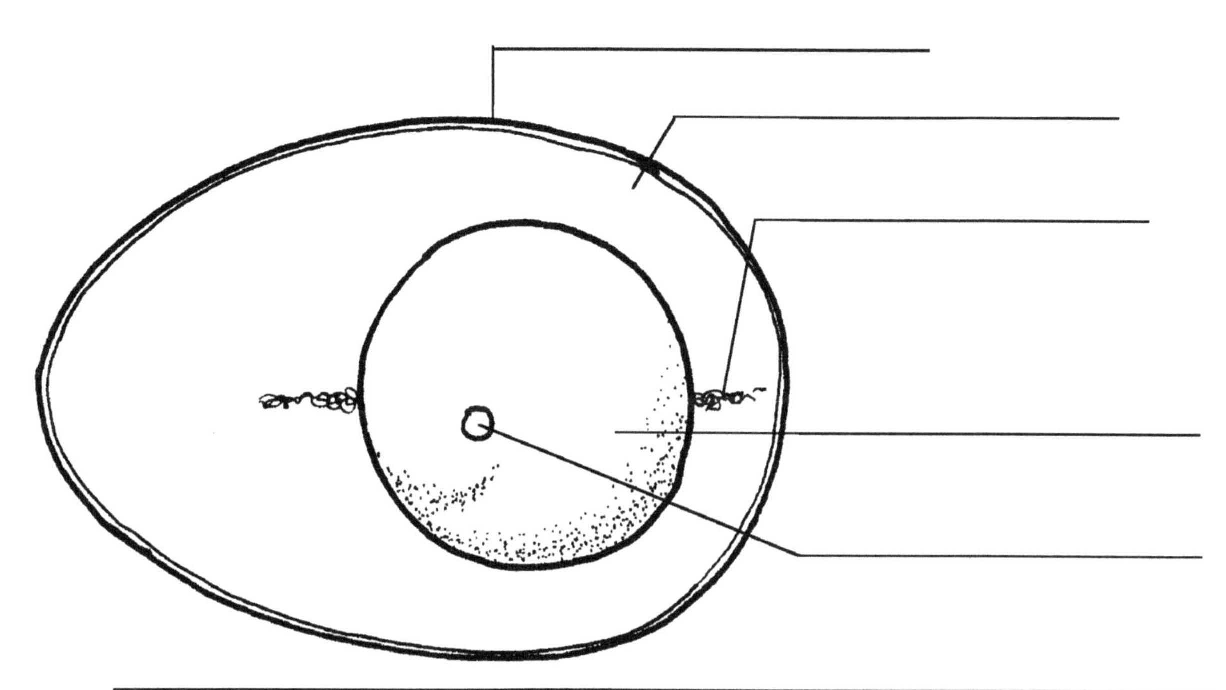

non-cell parts	shell	Protects the egg cell.	brown
	white	Cushions the egg cell.	yellow
	chalaza	Rope like strands that anchor the cell in place.	green
cell parts	yolk	Largest part of the cell. Food storage for the developing chick.	orange
	blastodisc	Contains most cell parts including the nucleus.	blue

1. In the box below sketch the actual cell parts of your egg.

Cell Lab #2: PLANT AND ANIMAL CELLS DIFFER - instructions

Material:

copy of lab sheet (1 page), pencil
6 oz. box light colored gelatin (pineapple or lemon)
small, square pan (to hold 2 cups of gelatin)
round bowl (to hold another 2 cups of gelatin)
20 or so green grapes (for chloroplasts), cut into half lengthwise
2 orange slices or 2 whole strawberries (for the nucleus of each cell)

Aloud: Even when you can find a cell big enough to see, you really can't see its parts very well without a microscope. Today we are going to make a model of a plant cell and a model of an animal cell to show how they are different and how they might look if you were shrunk down many sizes. Maybe for dinner you can sink your teeth into science.

Procedure: Lab Day:

1. *Do the "BEFORE THE LAB" section of the lab sheet.*
2. *Prepare gelatin as instructed on the box. Pour 2 cups into the square pan and 2 cups into the round pan.*
3. *After gelatin is mostly set, have students determine which pan represents the plant cell (the square one). Add the grapes (chloroplasts) to the plant cell. Stir them around into the gelatin.*
4. *Add one orange slice or strawberry to each bowl, pushing it into the middle of each one.*
5. *Complete lab. If desired, label the gelatin cell parts with flags on toothpicks. Make sure to include a label for the chloroplast, nucleus and cell membrane.*

Possible Answers:

1. *The rectangular one is the plant cell. The round one is the animal cell.*
2. *Chloroplasts go into the plant cell.*
5. *A plant cell: is rectangular and has chloroplasts.*

For More Lab Fun:

1. *Find other things to make cell models out of. How about Legos or K'Nex?*
2. *Act out the different parts of the cell. The cell membrane keeps out intruders (bacteria, toxins etc.) and keeps the cell together, the nucleus is the control center (boss) of the cell and the chloroplasts make food.*
3. *To help children understand the function of the cell parts, let them call the family boss (dad or mom) "nucleus" for a day. Maybe you have a dog that keeps out intruders. Let him be called the "cell membrane" for the day. Whoever is usually the cook can be called "chloroplast" for the day. Encourage your children to feed the "cell membrane" and help the "chloroplast" get dinner ready. Don't forget, when the "nucleus" tell you to clean your room, you better do it!*

Cell Lab #2: PLANT AND ANIMAL CELLS DIFFER - Instructions

[illegible]

NAME ______________________ DATE ______________

Cell Lab #2: PLANT AND ANIMAL CELLS DIFFER

BEFORE THE LAB:

1. Label the shapes below either "Animal Cell" or "Plant Cell."

2. In which cell will you put the chloroplasts (grapes)? ______________

AFTER MAKING THE CELLS: 3. Draw the cells that you made.

4. Label the cell membrane, nucleus and chloroplasts.
   The first pointer is drawn to show you how.

______________________ CELL

____

______________________ CELL

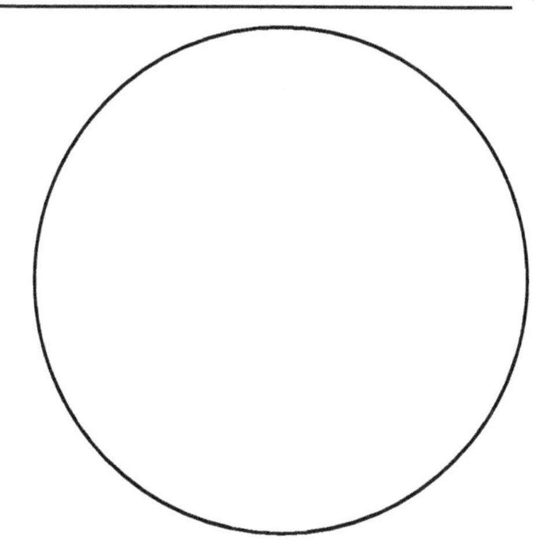

5. I can tell a plant cell from an animal cell because a plant cell: ______________

____

and ____

THE HUMAN BODY

by

HUMAN BODY FOLDER - instructions

Material:

copy of The Human Body divider (page 41), pencil
12" x 12" colored cardstock (I like scrapbooking paper as it lasts longer and doesn't fade)
12" x 18" construction paper or good, thick drawing paper
3 hole punch
glue gel (Elmer's School glue Gel- We call it blue glue because it is)
stapler scissors yarn
colored pencils

Aloud: We are going to start studying the human body- your body. Every part in your body helps you to stay alive. We will be learning what many of these parts do and how they work together. Can you imagine riding a bike if the wheels were missing? What if it had no pedals? Just like the different parts of a bike, your body parts must work together. If just one part is missing, or not doing its job, your body can stop working for you. Groups of body parts that help do a job together are called body systems. One body system keeps you fed, another keeps your blood moving. Before we learn about these systems, we are going to put together a folder to store all of your human body information in. It will be a folder of how your own body works.

Procedure:

Lab Day:

1. Decorate and color "The Human Body" divider page. Add your name to it. Glue it to the 12" x 12" page to make the front cover for your folder.
2. Fold up the bottom 6" of the 12" x 18" piece of paper so that you form a 12" x 12" page with a pocket at the bottom. Staple the right side closed. The pocket will be inside the folder.
3. With the 12" x 12" cover in front and the pocket inside, three hole punch the left side of both pages. Tie yarn in the holes to hold the folder together.
4. As you do the human body units, add any pages to your folder that you wish. Lab pages can be tied in with the yarn, paper models (like a muscle model and skeleton "puzzle" you will be making) can be put into the pocket.
5. Have fun learning all about how your body works!

NAME ________________________ DATE ________________

For my notebook

SKELETAL AND MUSCULAR SYSTEMS

Why aren't you just a gooey blob like a sea jelly? Because you have bones. These bones protect delicate organs like your heart, brain and lungs and they support you so that you can stand up and move. Did you know that most animals don't have any bones at all? Insects, worms, slugs, sea jellies and many others have no bones. Run your hands around your ribs. You can follow them from your vertebrae or backbones all the way around to the front by your chest. All of your ribs together form your rib cage. Why do you think they call it a "cage"? Feel the bony plate where your ribs meet in front. This is your sternum (**str**-nuhm) and it protects some very important organs inside of you. Bones are connected to each other in remarkable ways that allow us to move around, but without muscles to pull our bones back and forth we would be as still as a puppet with no strings. Bend your arm at the elbow. When you do this, the muscle on the front of your upper arm pulls to become shorter. When you straighten your arm again, the muscle in *back* pulls and shortens. These muscles that move you around aren't the only muscles in your body. Your heart is one special kind of muscle but there are many others. Muscles help with everything from breathing to digesting food to moving your eyes.

So, with bones to hold you up and muscles to get you moving you can go places no sea jelly could ever imagine.

Skeletal System Lab: MY SKELETON HOLDS ME UP - instructions

Material:

copy of lab sheet (1 page)
scissors
glue gel (Elmer's School Glue Gel)
1 - 8 1/2" x 11" black (or other color of choice) construction paper
white or gold pen that will mark on black paper

Aloud: Your skeleton is a framework of bones that gives you shape and holds you up. It also protects delicate organs from bumps and bruises. Bones have funny names but they aren't really hard ones to learn. You probably already know that your brain is protected by your skull. You may even know that hips are wide bones called the pelvis. When you hit your arm just right it sends an icky tingle down your arm doesn't it? So it isn't very humorous to hit your humerus, is it? Your femurs are the longest bones in your body. Can you find which ones those are? Have fun cutting out and gluing your skeleton puzzle together and as you go, think about which part of your body each bone is in.

Procedure:

Lab Day:

1. Cut around each section of bones. The black around each one gives little hands a little room for error.
2. Lay the bones on the black paper where you think they should go. Check arms and legs, feet and hands on the sample in the corner of the lab sheet.
3. Once you have everything where it belongs, remove and glue the pieces down one at a time.
4. If you wish, use a white or gold pen to label the major bones.

Possible Answers:

See sample on lab sheet for correct bone placement

Conclusion / Discussion:

1. Without bones animals can't get very tall, or must live in water so the water can hold them up. Sea jellies are tall but they float in water. On land the tallest invertebrate (boneless animal) is probably an insect. Doesn't compare to human height, does it?
2. How do we know there will never really be monstrously large spiders and scorpions like we see in science fiction movies?

For More Lab Fun:

1. Play Simon Says using the names of bones. You can pat your skull or touch your patellas, tickle your ribs or thump your sternum.
2. Next time you pull apart a chicken or turkey for dinner, take a good look at the bones. Compare them to human bones.
3. Older children may want to identify and label the following bones as well

   lower arm bones are the radius and ulna
   lower leg bones are the tibia and fibula
   fingers and toes are all phalanges
   your jaw is your mandible
   your shoulder blades are called scapula
   your collar bone is your clavicle

Skeletal System Lab: MY SKELETON HOLDS ME UP

Humerus (hue-mr-us)

Ribs

Pelvis

Vertebrae
(vr-tuh-bray)

Skull

Femur

Patella
(puh-tel-uh)

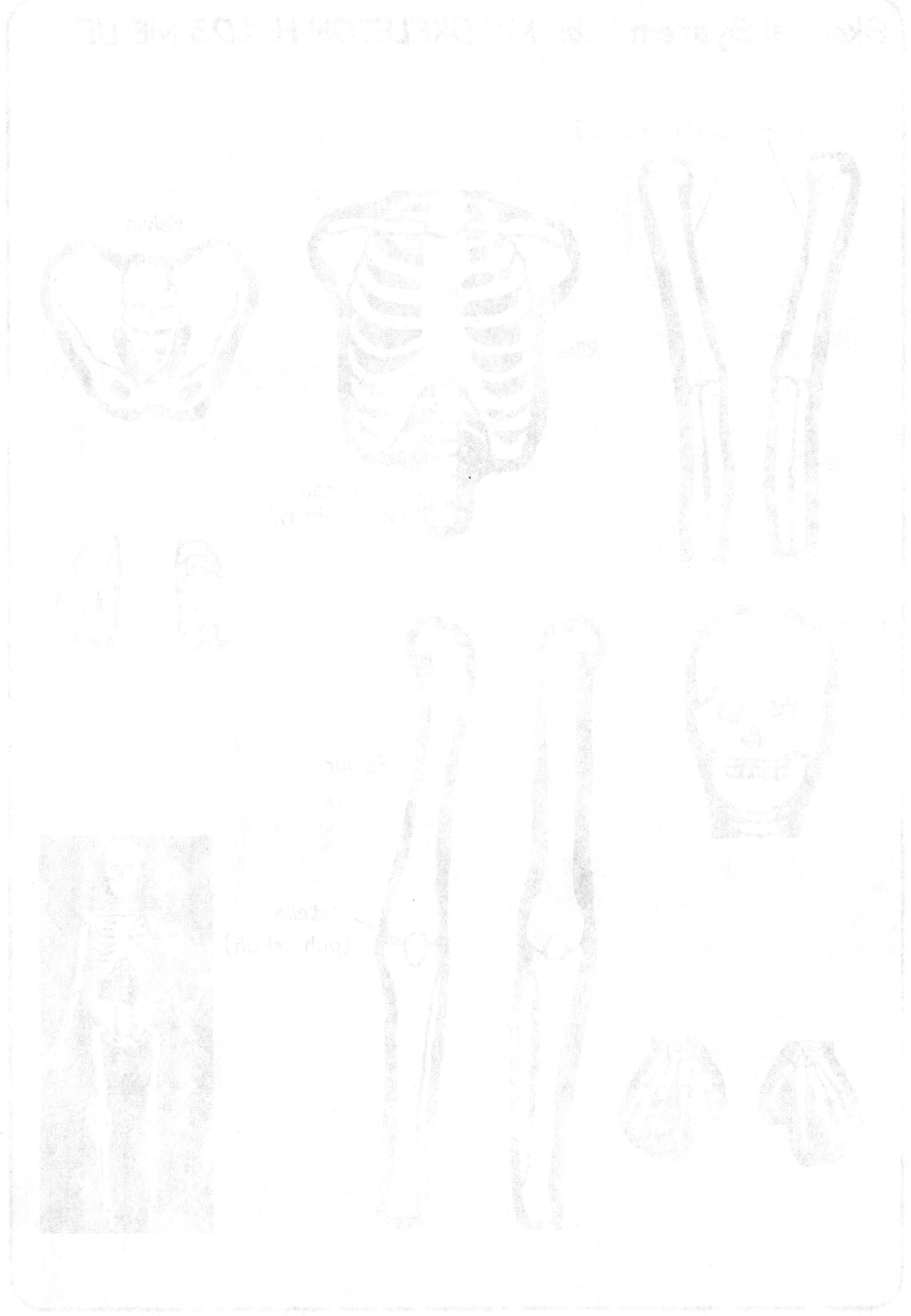

Muscle Lab: MUSCLES AREN'T PUSHY - instructions

Material:

copy of lab sheet (3 pages), pencil
1 - tagboard or very heavy copy paper - 8 1/2" x 11"
pair of scissors
hole punch
1 metal fastener
string or thread 70 cm. long (about 28") cut in half
tape ruler

Aloud: Bones hold us up but without muscles to move those bones we'd be a lot like a tree -stuck in one spot. Muscles pull to move bones but they can't push. Because of this, muscles have to work in twos. To bend your arm, one muscle pulls and then to straighten your arm back out, a muscle on the other side has to pull. We're going to make a model of an arm and show how two muscles work together to move it.

Procedure:

Lab Day: Make Your Arm Model

1. Trace or photocopy pattern of upper and lower arm onto thick paper. Cut out.
2. Punch holes at circles and at center of X's. You will have 3 holes in piece Q and 1 hole in piece Z.
3. Place pieces Q and Z as indicated in top diagram on page 56. Push metal fastener into holes at the X's to secure pieces.
4. Thread one piece of string from back, through each hole at the top of Q and tape them down onto piece Z where indicated in bottom diagram.

Using Your Model

5. With arm in position shown in diagrams, pull slack from strings then measure and record length from end of hole to start of tape. Record on line labeled "Beginning."
6. Carefully pull string #1 until arm can't move any further. Again, measure both strings and record on chart. Compare lengths with those from the beginning to fill in all answers in the row for "String #1 pulled."
7. Pull string #2. Repeat measurements and record.
8. Finish filling in chart for "String #2 pulled."
9. Answer lab questions.

Possible Answers:

1. Lower arm and hand or forearm.
2. Upper arm
3. Muscles
4. 2
5. Front

Conclusion / Discussion:

- Help children notice that as a muscle pulls it is actually becoming shorter. When the opposing muscle (the one that does the opposite motion) pulls, the original muscle relaxes and gets longer.

NAME ______________________________ DATE ____________________

Muscle Lab: MUSCLES AREN'T PUSHY- pg. 1

Trace the parts below onto tagboard. Cut out and follow the directions to assemble your arm model.

Q

X

X

Z

NAME ______________________ DATE ______________________

Muscle Lab: MUSCLES AREN'T PUSHY- pg. 2

	Length of String #1	Length of String #2	String #1 gets (Shorter or longer)	String #2 gets (Shorter or longer)	Arm (bends or straightens)
Beginning	cm.	cm.	______	______	______
Pull String #1	cm.	cm.			
Pull String #2	cm.	cm.			

1. What part of your body is part Z supposed to be? ______________________
2. What part is part Q supposed to be? ______________________
3. The strings are supposed to be ______________
4. How many muscles does it take to move an arm up and down? ______
5. When your elbow bends which muscle is getting shorter, the muscle in the front or the one in the back of your arm? __________

Muscles Lab: MUSCLES AREN'T PUSHY- pg. 3

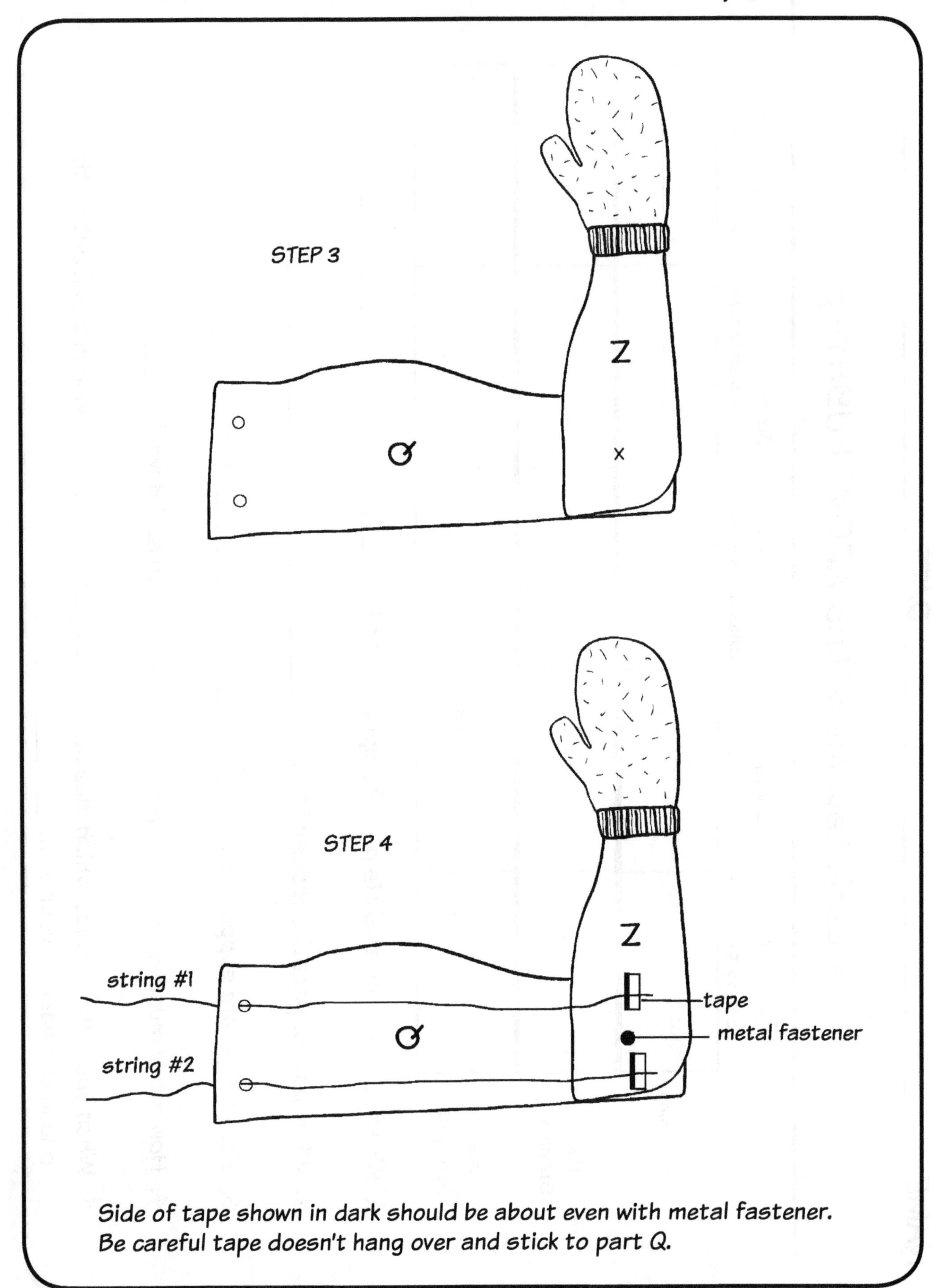

Side of tape shown in dark should be about even with metal fastener.
Be careful tape doesn't hang over and stick to part Q.

NAME ______________________ DATE ______________

For my notebook

CIRCULATORY SYSTEM - YOUR HEART AND BLOOD

You know that muscles are important body parts, but which muscle is the MOST important? Your heart would probably have to take that prize. Your heart is a muscle that pushes blood all through your body. This blood carries oxygen, food and water and it carries special cells that fight diseases. Blood travels through tubes called blood vessels. The vessels going away from your heart are called arteries (**ahr**-tuh-reez). The vessels going back to your heart are called veins (vaynz). How long are all these vessels? Well, if you could lay them all end to end they would go all the way around the world. Now, hold up your hand and make a fist. Your fist is about the size of your heart. Not very big for all the work it does, is it? Your heart may not be very big but it sure is powerful. Try to squeeze a tennis ball hard enough for it to squish in. That's how hard your heart squeezes and it does that between 80 and 100 times each minute. Now that is a powerful muscle! So, your circulatory system moves blood all through your body. The heart is the muscle that pushes the blood away through arteries and then back again through veins. Along the way, blood carries oxygen, food and water as well as cells that fight diseases. By drinking water, eating right and exercising regularly you can help keep this amazing system healthy for many years. Now that doesn't seem like too much to ask for something that works that hard for you, does it?

Circulatory System Lab #1: YOUR HEART RATE - instructions

Material:

copy of lab sheets (2 pages), pencil
watch with second hand
crayons- 6 colors

Aloud: Your heart works hard to supply your whole body with the food, water and oxygen it needs. Your body needs these things just like a car needs gasoline. Without fuel, your body will stop working. In fact, the harder your body works, the faster it uses up the fuel in it. When you are exercising hard, your heart pumps faster to hurry food, oxygen and water to all the parts of your body. When you slow down, so can your heart because your body doesn't need as much of these things. In this lab you will compare how fast your heart is working when you are resting with how fast it works when you are exercising. Let's see just how well your heart can tell how much work your body is doing.

Procedure:

Lab Day:

1. Fill in questions 1 - 4 on lab page 1.
2. PARENTS: It's might be easiest if you to take your child's pulse for him. Younger children often lose count easily.
3. Practice finding your (or your child's) pulse. One easy place to find a pulse is on the wrist. Using your index and middle fingers but NOT your thumb (which has it's own pulse) find the stiff ridge that runs down the inside of the wrist on the thumb side. Place two fingers on the thumb side of this ridge. Push down slightly and you should find a pulse.
4. Sit very still and quiet for a minute. Take pulse for 20 seconds. Continue with activities and checking pulse. Fill in beats per 20 seconds. For exercise #10 choose your own favorite. Depending on math level, either add pulse three times (for example, if you count 60 heartbeats in 20 seconds, you would add 60 + 60 + 60) or multiply pulse by 3 (60 x 3) to find pulse for one minute. Finish page one.
5. Fill in and color graph to show comparison between resting and working heart rates. Complete lab.

Possible Answers:

4. My heart pumps faster when I am working hard because my body needs more fuel when it is working.

Conclusion / Discussion:

1. How did the difficulty of the exercise compare to the heart rate? Why is this?

For More Lab Fun:

1. Take pulse every minute after exercising. See how long it takes pulse to return to normal. Graph the results.
2. Take a field trip to a blood bank. See how they screen donors and take blood. Learn about blood types.
3. Learn the basics of CPR.

NAME ______________________ DATE ______________________

Circulatory System Lab #1: YOUR HEART RATE - pg. 1

Of the activities below, I think:

1. ______________________ will be the easiest and
2. ______________________ will be the hardest.
3. I think my heart will beat fastest when I ______________________
4. I think my heart will beat slowest when I ______________________

My heart rate for 20 seconds of:

5. Sitting still = _______ beats X 3 = _________ beats each minute
6. Walking (1 minute) = _______ beats X 3 = _________ beats per minute
7. Jumping Jacks (10) = _______ beats X 3 = _________ beats per minute
8. Sit ups (10) = _______ beats X 3 = _________ beats per minute
9. Push ups (10) = _______ beats X 3 = _________ beats per minute
10. ____________ = _______ beats X 3 = _________ beats per minute

(exercise of my choice)

Circulatory System Lab #1: YOUR HEART RATE- pg. 2

Enter the heart rates you calculated onto the graph below.
Color in the graph using a different crayon for each activity.

	50	60	70	80	90	100	110	120	130	140	150	160	170	180	190	200	210	220	230	240	250	260
SITTING																						
WALKING																						
JUMPING JACKS																						
SIT UPS																						
PUSH UPS																						
MY OWN																						

HEART BEATS PER MINUTE

1. ________________ was the easiest for me to do. My heart beat slowest when I did ________________________

2. ________________ was the hardest for me to do. My heart beat fastest when I did ________________________

3. ________________ exercises that make me work hardest, make my heart pump faster.
   (The same or Different)

4. My heart pumps faster when I am working hard because ____

   ____

Circulatory System Lab #2: BLOOD MODEL - instructions

Material:

copy of lab sheets (2 pages), pencil
crayons- tan, blue, yellow, red
1 ½ cup or larger clear jar with wide mouth
stirring spoon
1/2 cup light Karo syrup (PLASMA) (**plaz-muh**)
1/2 cup Red Hots candies (RED BLOOD CELLS)
5 dry large lima beans (WHITE BLOOD CELLS)
1 tablespoon dry lentils or split yellow peas (PLATELETS) (**playt-lets**)

Aloud: Blood looks like a smooth, red liquid, but it's not. Blood is actually a straw colored liquid with tiny cells mixed in it. Your lab today will show you the main blood parts so you can learn what they are for and what they look like. Blood parts are so tiny you can't see them without a microscope so, after coloring the blood picture, we will be making a model of blood to show how the parts look close up.

Procedure:

Lab Day:

1. Color in the blood picture as indicated. Make sure to read all of the information on this sheet.
2. Measure out the amounts of Karo syrup, Red Hots, lima beans and lentils indicated above. Pour them into the jar, one at a time, calling them the blood parts they represent. For instance, don't say "Pour in the Karo syrup." Instead say, " Pour in the Plasma."
3. Mix gently with spoon and compare to real blood.

Possible Answers: (to page 66)

Red hots = red blood cells
Karo syrup = plasma
Dry lima beans = white blood cells
Dry lentils = platelets
Plasma keeps things moving and transports food
Red blood cells carry oxygen
White blood cells fight diseases by eating harmful bacteria
Platelets plug up wounds to help clot the blood

Conclusion / Discussion:

- Discuss the following: Is blood all red? (No). What part of blood gives it its red color? (red blood cells). How many major parts make up blood? (4).

For More Lab Fun:

1. Look at the underside of your tongue in the mirror. Can you see where the blood has no oxygen (blue veins) and where the blood has oxygen (red arteries)?
2. Make a round cake. Frost it with yellow frosting (plasma). Decorate with white jelly beans (white blood cells), red M & Ms (red blood cells) and silver cake decorating balls (platelets). Put on a vampire cape and share your blood with your friends.

For Your Information:

1. Blood is red because red blood cells have iron in them. Iron mixed with oxygen becomes red. On the way back to the heart, after giving up its oxygen, blood is a bluish color. Insects and other animals have different colored blood because their blood cells aren't iron-based.
2. Arteries carry blood away from the heart so they usually contain red, oxygenated blood. Veins carry blood back to the heart so they usually carry blood with no oxygen.

NAME ______________________________ DATE ________________

Circulatory System Lab #2: BLOOD MODEL- pg. 1

This is what a drop of blood looks like under a microscope. Read the clues below and color the parts of the blood as indicated. Next, read the jobs each blood part performs. Write the name of that blood part.

<u>White Blood Cells</u>: I am the biggest cell in your blood. I have a nucleus that is often split into 2 or 3 parts. Color me tan.
<u>Plasma</u>: I am mostly water. I make up about half of your blood. Color me yellow.
<u>Platelets</u>: I am the tiniest part in your blood. Color me blue.
<u>Red Blood Cells</u>: I am shaped like a round candy, dented in on both sides. There are more of me than any other blood cells. You can color me red.

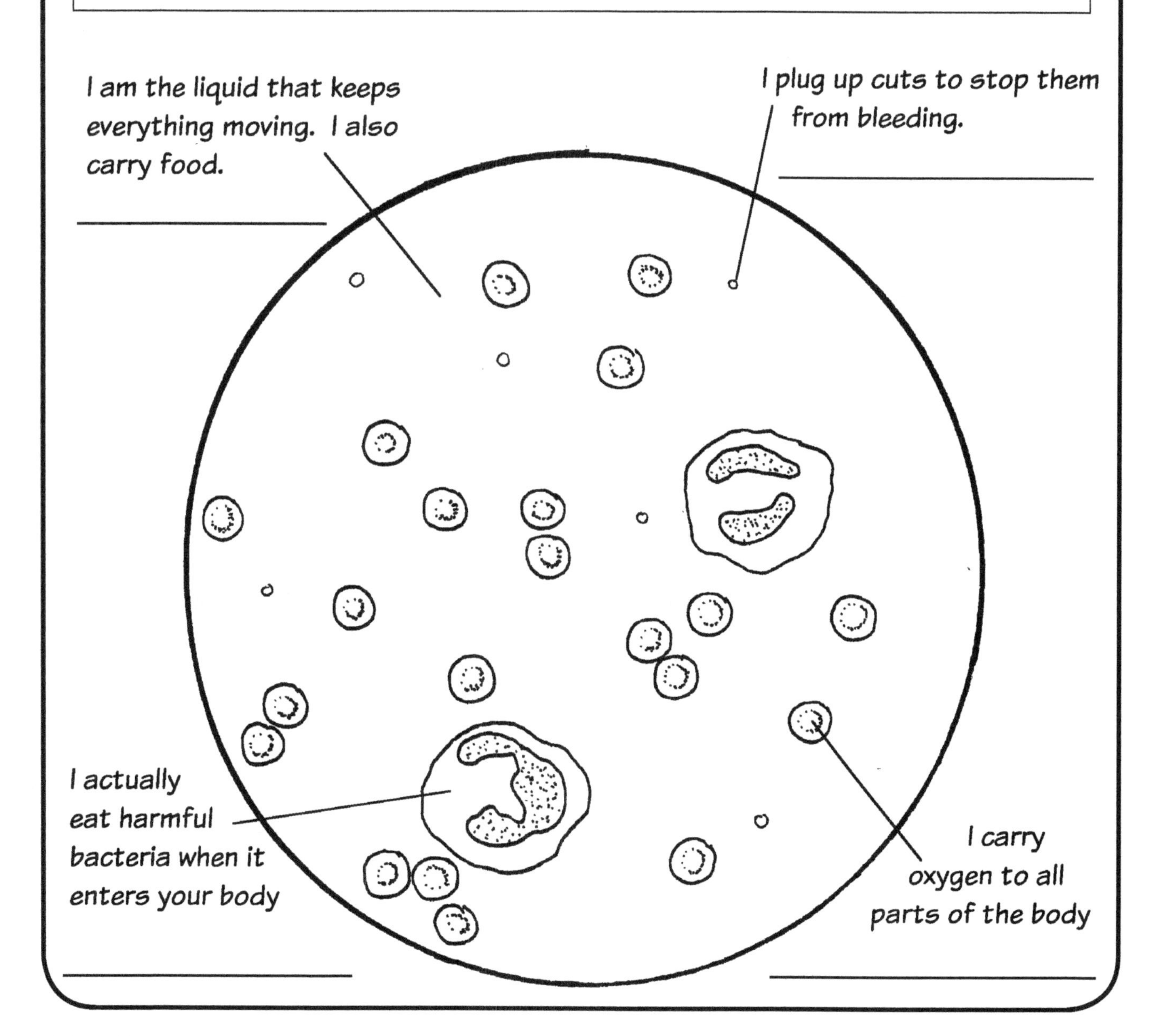
I am the liquid that keeps everything moving. I also carry food.
I plug up cuts to stop them from bleeding.
I actually eat harmful bacteria when it enters your body
I carry oxygen to all parts of the body

Circulatory System Lab #2: BLOOD MODEL- pg. 2

Draw your blood model in the square below.

A person's heart weighs about 1 pound.
An elephant's heart weighs about 44 pounds!

What am I and what am I delivering?

<u>Which blood part does each item represent?</u>

Red hots candies = ____________________

Karo syrup = ____________________

Dry Lima Beans = ____________________

Dry Lentils = ____________________

<u>In the spaces below, write the job done by each part of your blood.</u>

Plasma ____

Red Blood Cells ______________________________________

White Blood Cells ____________________________________

Platelets ____

NAME ______________________________ DATE ___________________

For my notebook

YOUR RESPIRATORY SYSTEM BREATHES FOR YOU

What surrounds you all the times but you cannot see it, feel it, hear it, taste it or smell it? Air! There is nothing more important to your body than air. Air is a mixture of many things but the oxygen in air is what your body needs most. Oxygen is a gas that your body combines with the food you have eaten to make energy. When you run your body needs more energy, doesn't it? Because of this, when you exercise, you breathe faster to get more oxygen to make more energy. When you slow down, your breathing can slow down too. When you inhale or breathe in, air goes in your nose or mouth, down a tube called a trachea (**tray**-kee-uh) and into your 2 lungs. Your lungs are big bags made of a bunch of tiny bags that fill with air. Blood is pumped in tubes all around these tiny air sacs. Oxygen from the air goes into the blood and carbon dioxide, a waste your body makes, comes from your blood and goes into the air sacs. When you breathe out, the carbon dioxide leaves your body. In with the good and out with the bad. It's a great system and it shows how all of our body systems work together to keep us alive.

Your respiratory system breathes in air, passes the oxygen to your circulatory system—which combines it with food from your digestive system to make energy for your muscular system. Whew! And I thought breathing was so simple!

Respiratory System Lab #1: BREATHING RATE - instructions

Material:

copy of lab sheets (2 pages), pencil
completed "Circulatory System lab- YOUR HEART RATE" page 62 (for reference)
watch with second hand
crayons- 6 colors

Aloud: When you breathe your body takes in oxygen. This oxygen is carried to all the cells in your body. Oxygen is needed by your cells to turn food into fuel. Without this fuel your body wouldn't be able to grow and do work. If oxygen is needed for your cells to do work, how do you think your breathing will change when your body is working harder than normal? We are going to do a lab to find out how hard work and exercise effect the way you breathe.

Procedure:

Lab Day:

1. Fill in questions 1 - 4 on lab page 1.
2. PARENTS: It is probably easiest if you count your child's breaths. Younger children often lose count easily.
3. Sit very still and quiet for a minute. Count the breaths you take for 30 seconds. Continue with activities and checking breathing rates. Fill in breaths per 30 seconds. For exercise #10 choose your own favorite exercise. Depending on math level, either add breaths two times (for example, if you take 20 breaths in 30 seconds you would add 20 + 20) or multiply breaths by 2 (20 x 2) to find breathing rate for one minute. Finish page one.
4. Fill in and color graph to show comparison between resting and working breathing rates. Complete lab. Refer to YOUR HEART RATE lab page 62 to answer question #3 (hardest exercise for my heart)

Possible Answers:

#4. My heart and lungs work together to take oxygen from outside the body and move it to all the cells in my body.

Conclusion / Discussion;

1. Discuss results. How did the difficulty of the exercise compare to the breathing and heart rates? Why is this?
2. Discuss how the circulatory, respiratory and digestive systems are interconnected. They all do their part to provide the body with the chemicals it needs to survive.

For More Lab Fun:

1. Learn the basics of the Heimlich maneuver.
2. Develop an exercise routine that conditions your muscles as well as your heart and lungs.

NAME ______________________ DATE ______________________

Respiratory System Lab #1: BREATHING RATE- pg. 1

<u>Of the activities below, I think:</u>

1. ______________________ will be the easiest and
2. ______________________ will be the hardest.
3. I think my breathing will be slowest when I ______________
4. I think my breathing will be fastest when I ______________

<u>My breathing rate for 30 seconds of:</u>

5. Sitting still = _______ breaths X 2 = ________ breaths per minute
6. Walking (1 minute) = _______ breaths X 2 = ________ breaths per minute
7. Jumping Jacks (10) = _______ breaths X 2 = ________ breaths per minute
8. Sit ups (10) = _______ breaths X 2 = ________ breaths per minute
9. Push ups (10) = _______ breaths X 2 = ________ breaths per minute
10. ____________ = _______ breaths X 2 = ________ breaths per minute

(exercise of my choice)

Respiratory System Lab #1: BREATHING RATE- pg. 2

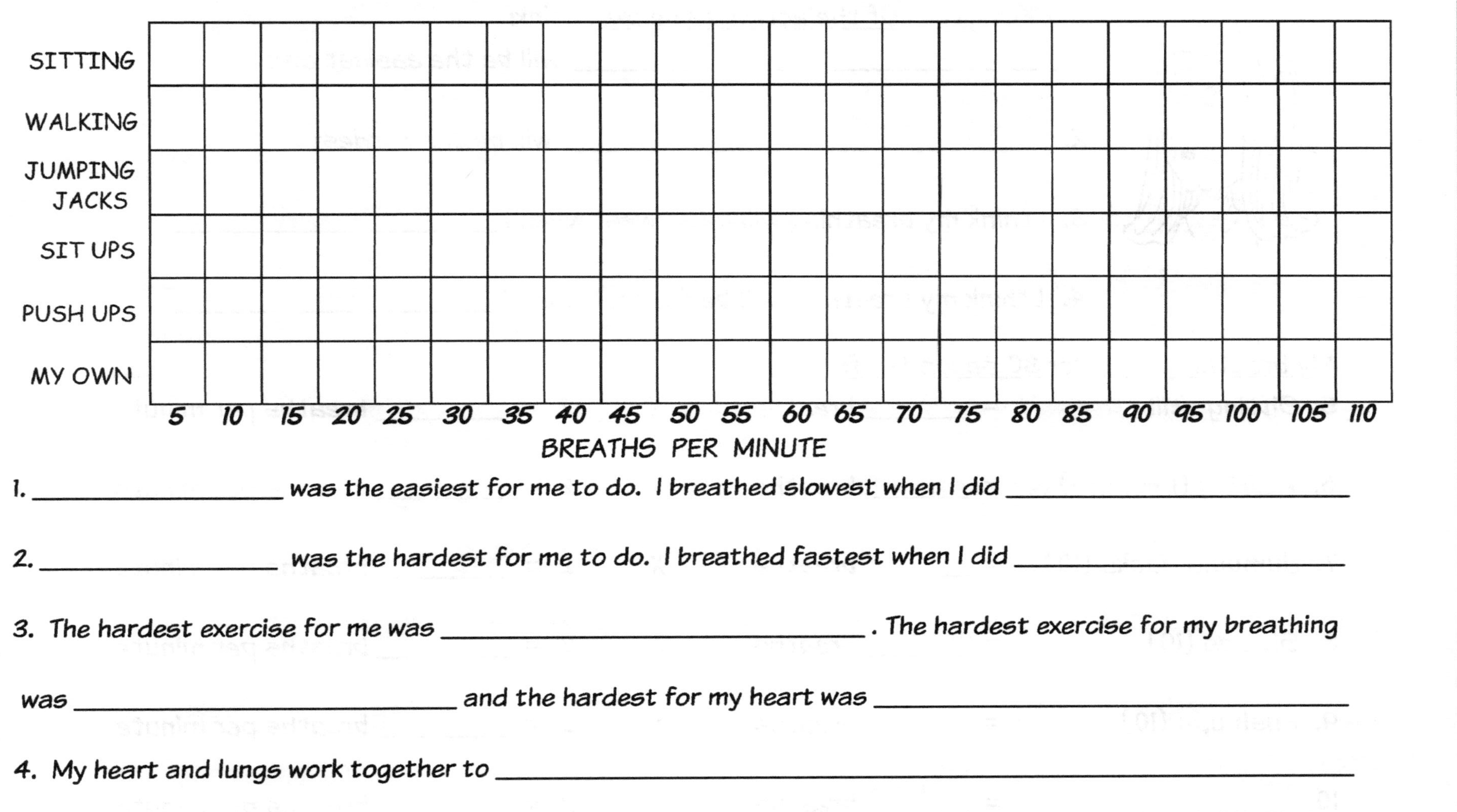

Enter the breathing rates you calculated onto the graph below.
Color in the graph using a different crayon for each activity.

	5	10	15	20	25	30	35	40	45	50	55	60	65	70	75	80	85	90	95	100	105	110
SITTING																						
WALKING																						
JUMPING JACKS																						
SIT UPS																						
PUSH UPS																						
MY OWN																						

BREATHS PER MINUTE

1. ________________ was the easiest for me to do. I breathed slowest when I did ________________

2. ________________ was the hardest for me to do. I breathed fastest when I did ________________

3. The hardest exercise for me was ________________ . The hardest exercise for my breathing was ________________ and the hardest for my heart was ________________

4. My heart and lungs work together to ________________

____

Respiratory System Lab #2: I NEED OXYGEN - instructions

Material:

copy of lab sheet (1 page), pencil
4 pieces of 8 1/2" x 11" paper or paper plates
long tube (like vacuum cleaner tube or wrapping paper tube) label the tube "trachea"
red and blue math counters or chips about 10 each color
colored pencils- various colors
a helper/teacher (acting as the giant)

Aloud: Today you get to be a red blood cell (RBC) carrying oxygen through the body of a giant. You are going to start your journey in the giant's heart. From there you will go to the lungs, pick up some oxygen and go back to the heart. The giant's heart will pump hard enough to get you all the way to the giant's foot. Because the giant is walking, his foot will be using up oxygen and making carbon dioxide- a waste product. You will need to trade your oxygen for the carbon dioxide, take it to the heart so it can send you back to the lungs. Once in the lungs the giant will breathe out, sending the carbon dioxide out of his body. You will pick up more oxygen and go back around again. It's a long way to go for a little blood cell, like yourself, but you know it is a job that must be done for the giant to live.

Procedure:

Lab Day:

1. Using your highly developed artistic abilities, draw a big nose on one paper, a heart on another, a lung on another and a foot on another. Label each one. Color them, if you would like.
2. TEACHER: Lay out the papers as shown on the Respiratory Setup on back of this page (page 74). Lay the tube between the nose and the lungs. Spread the others out so that the entire giant is about 10 feet tall. The tube is the trachea. Place 5 blue chips on the foot plate.
3. GIANT: (inhales, bringing oxygen into his lungs)- slide 1 red chip down the trachea onto the paper lungs.
4. RBC: 1) Start in the heart 2) Go to the lungs to pick up one oxygen 3) Back to the heart 4) Out the heart and down to the foot. At the foot, trade your oxygen for one carbon dioxide there 5) Head back to the heart with the carbon dioxide 6) And finally back to the lungs to get rid of your carbon dioxide.
5. GIANT: (exhale) slide the blue chip up towards the nose so it can leave the body. Inhale again, sliding another red chip onto the lung.
6. Repeat as often as you would like to get children to understand the pathways of the blood, oxygen and carbon dioxide.
7. Have children complete the lab sheet, using the appropriate colored pencil (red or blue) to color in the path of the oxygen and carbon dioxide.

Possible Answers:

SEE BACK OF PAGE FOR SETUP AND ANSWERS

For More Lab Fun:

1. Continue with the giant breathing but start with all 10 blue chips on the plate and "breathe" faster, sliding the red chips onto the plate more quickly. Ask your child what he thinks is happening (the giant is exercising and using up more oxygen). The blood has to get moving to keep up with the oxygen demands. Go faster and slower, asking what might be going on (The giant could be riding a bike, watching a scary movie, sleeping etc.)

Respiratory System Lab #2: I NEED OXYGEN - instructions pg. 2

SETUP:

Set up your giant as shown below. Make sure the parts are fairly far apart (about 10' total). Arrows are the answers for questions #1 and #2 of the lab. Don't show them to the students!

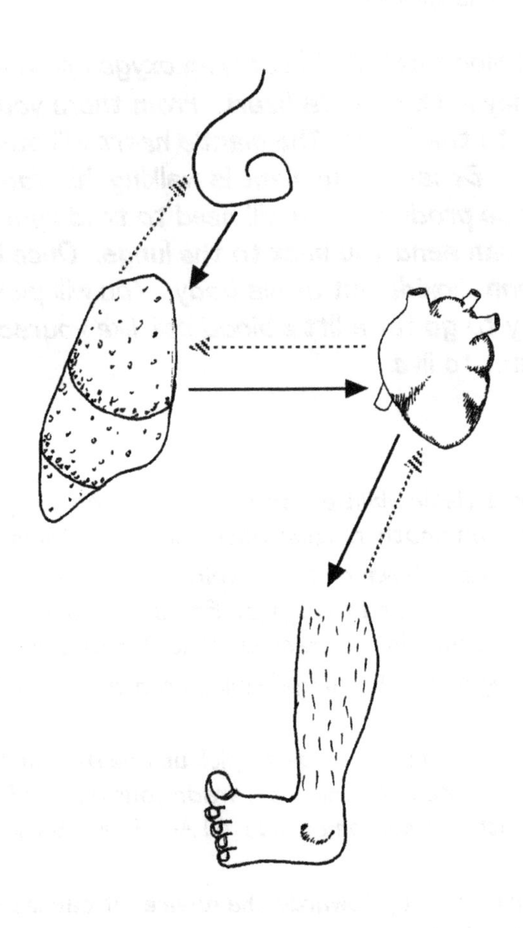

POSSIBLE ANSWERS:

#1 (solid lines = red = oxygen) Note: Oxygen makes the hemoglobin in blood turn red so oxygenated blood is red.

#2 (dashed lines = blue = carbon dioxide) Note: De-oxygenated blood (blood without oxygen) has a bluish color.

#3 When you exercise your body is using more oxygen and producing more carbon dioxide so RBCs have to deliver them faster.

#4 Your RBCs are moving faster because your heart is pumping faster. You can feel your heart, breathing and pulse speed up.

CONCLUSION / DISCUSSION:

1. Tell children that oxygen makes the blood red and when there isn't oxygen in it blood looks blue. Have them look at the underside of their tongues to see blue veins. To show red, oxygenated blood, in semidarkness, shine a flashlight through a finger or into your mouth, with your mouth closed around the light. You'll be surprised at how red your cheeks are with light shining through them. It's fun!

NAME ______________________ DATE ______________

Respiratory System Lab #2: I NEED OXYGEN

1. With a red pencil, use arrows to show the path oxygen takes from where it enters the body until it reaches the foot.
2. With a blue pencil, use arrows to show the return journey made by carbon dioxide.

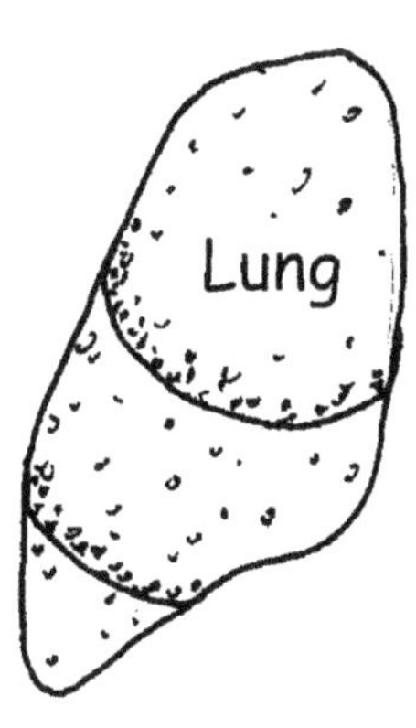

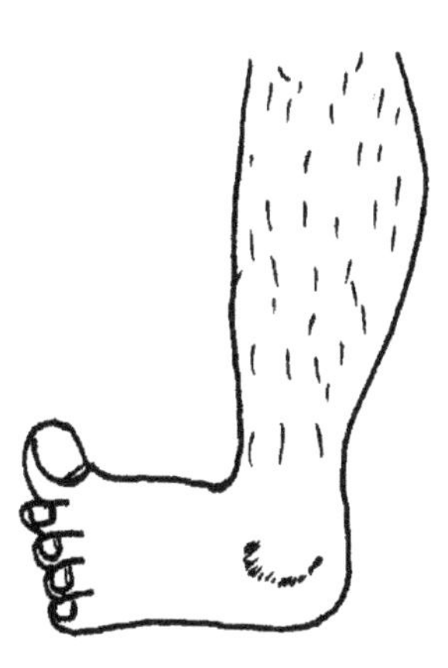

3. Why do your RBCs have to move faster when you are exercising?

____

4. How can you tell your RBCs are moving faster? Can you feel them?

____

NAME ______________________________ DATE ____________________

For my notebook

THE DIGESTIVE SYSTEM FUELS YOUR CELLS

What is your favorite fruit? How about your favorite vegetable? Did you know that the food you eat is used to make new cells so your body can grow and repair itself? It also provides energy to your body cells to do work. Do you remember how your heart and breathing get faster as you work harder? That's because the cells all over your body need more oxygen and food as you work harder. Your blood delivers that oxygen and food. When you eat, you know that food goes to your stomach, but to your brain, skin and every other part of your body? Yes! Food, water and oxygen go to all parts of your body to make new cells and to keep the cells you already have healthy. Have you heard the saying "You are what you eat"? It's really true. When you eat a handful of peanuts, new cells all over your body will be made of tiny parts from those peanuts. It takes chewing, sloshing, mixing and a lot of time to turn those peanuts into the parts needed to make cells. In fact, food can stay inside your body for 2 days. Your digestive system is more than just a stomach. It starts at your teeth, goes to your esophagus (ih-**sohf**-uh-guhs), stomach, small and large intestines (in-**tehs**-tihns) and ends at your anus, (**ay**-nuhs) where the waste leaves your body.

So, let's find out what it takes to make a bunch of broccoli into a batch of body cells.

Digestive System Lab #1: DIGESTIVE DIAGRAM - instructions

Material:

copy of lab sheet (1 page), pencil
Digestive System Notebook page (for reference)
colored pencils (blue, green, yellow, red, brown)

Aloud: It takes a lot of work to make food into energy for your body. We learned what some of the parts of the digestive system are called. Now let's color in a diagram of the digestive system to learn what those parts look like.

Procedure:

Lab Day:

1. Follow the instructions on the lab sheet. Label each part as numbered and then color in as indicated.
2. Follow the instructions at the bottom of the page.

For Your Information:

See diagram below for answers

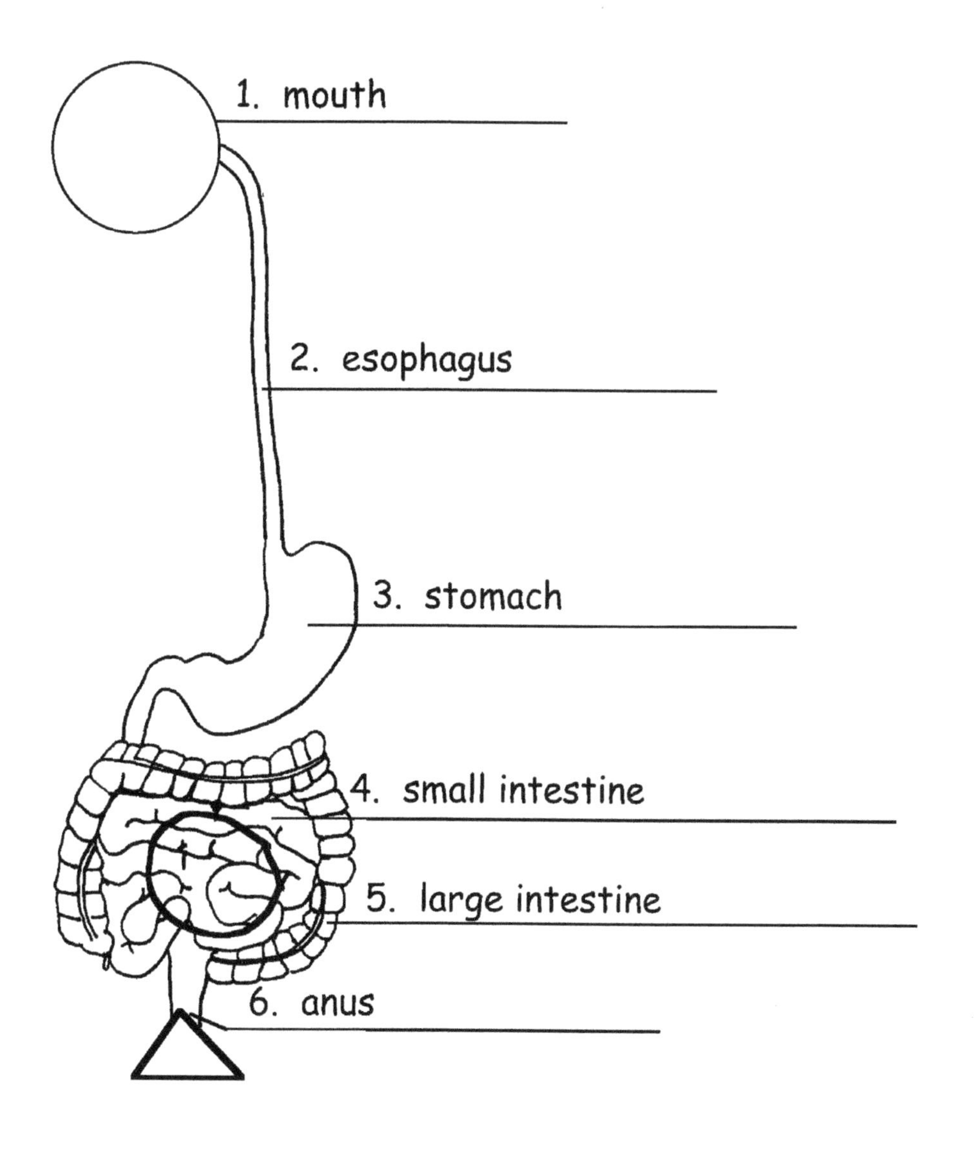

NAME ______________________ DATE ______________

Digestive System Lab #1: DIGESTIVE DIAGRAM

Use the key to label the parts of the digestive system and then color them in with the colors indicated.

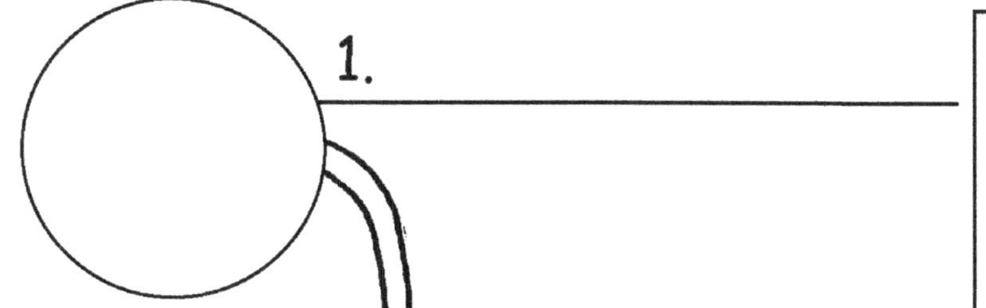

1. mouth - leave uncolored
2. esophagus - blue
3. stomach - green
4. small intestines - yellow
5. large intestines - red
6. anus - brown

2. ______________________

3. ______________________

4. ______________________

5. ______________________

6. ______________________

Draw silly teeth in the circle to show where food is first broken down.
Draw a circle on a coiled part of the digestive system.
Draw a triangle showing where undigested food and waste leave the body.

Digestive System Lab #2: FOOD TRAVELS FAR – instructions

Material:

copy of lab sheet (1 page), pencil
yarn- about 22' long
yard stick
5- 3" x 5" recipe cards
stapler

Aloud: "Mmmm. Great muffin. Thank you." The food you eat takes a longer journey through your body than you might expect. After you finish chewing up that muffin, your tongue pushes it back to be swallowed and it goes into a long tube called the esophagus. The esophagus squeezes the now smashed up muffin down into your stomach where it is mixed and mashed some more. After about 3 hours the muffin will leave your stomach and enter your small intestines. From there it will go on to the large intestines before leaving your body. Let's find out just how long your digestive tract is.

Procedure:

Lab Day:

1. Complete the "Hypothesis" section of the lab.
2. From one end of the yarn, mark off 3 inches. This is the length of the average child's mouth. Tie a knot in the yarn at that mark and label the index card "Mouth – 3 inches." Staple card to that section of the yarn.
3. From the first knot, measure and mark off 10 inches for the esophagus. Tie a knot and label the next index card "Esophagus – 10 inches."
4. Food dumps from the esophagus into the stomach. Measure 6 inches for the length of a child's stomach. Knot the yarn and label this section "Stomach – 6 inches."
5. The small intestine is the shocker. Measure off 15 feet for the small intestines. Food can spend about 3 hours in the small intestines where digestion is completed and from which the nutrients from the food can be absorbed. Knot and label this section "Small intestine – 15 feet."
6. Finally, measure off an additional 4 feet for the large intestines where water will be absorbed into the body. Cut the yarn at this point and label this section "Large intestine – 4 feet."
7. Stretch out your entire digestive system. Were you surprised at how long it is?
8. Enter actual lengths on #4. For #5, add together lengths, rounding to the nearest foot (total = 21').
9. Subtract your guess from the actual length to find how far off you were. Discuss the difference. Were you surprised? Were you close?
10. For #7 divide length of yarn by height of child OR have child lay next to yarn, mark where his head reached, have child move up yarn to there and repeat until he has figured out how many body lengths long his yarn (digestive tract) is.
11. For #8: At three meals a day, in three days, your child will have eaten 3 x 3 = 9 meals before his last one is all the way through his digestive tract.

Possible Answers:

Answers provided in the directions.

Conclusion / Discussion:

1. Discuss how many steps and how far food travels before it is completely digested.
2. Discuss the fact that at any one time a person has 9 past meals somewhere within their digestive tract. It's amazing we EVER feel hungry!

For More Lab Fun:

1. Eat an apple (or anything else). Throughout the day, announce the times it reaches each new point as follows: Esophagus in 15 seconds, stomach in 10 more seconds, small intestine in 3 hours, large intestine in 3 more hours and anus (exit) in about 2 days. You will need a calendar for the last one. In 2 days I want you to call out, "Hey everybody! The apple's done!"

NAME ______________________ DATE ______________

Digestive System Lab #2: FOOD TRAVELS FAR

HYPOTHESIS (MY BEST GUESS):

1. My height to the nearest foot (measure) ____________
2. The length of my digestive tract (estimate) ____________ feet
3. The time it will take a muffin to be digested ________________

TEST:

4. Length of the:

mouth + ________________

esophagus + ________________

stomach + ________________

small intestine + ________________

large intestines + ________________

5. Length of the entire digestive tract rounded to the nearest foot = ________________

RESULTS:

6. Actual length - My guess = Amount I was off

__________ - __________ = ________________

CONCLUSION:

7. I would have to lay ______ of me down, end to end to make the length of my digestive tract.

8. My muffin will be finished digesting in about 3 days. I will have eaten _______ more meals by that time.

NAME ______________________________ DATE ___________________

For my notebook

MY NERVOUS SYSTEM IS IN CONTROL

Can you pat your head and rub your tummy at the same time? Well, that may be a bit tricky but I bet you can walk across the floor and eat an apple at the same time. Your brain is doing an incredible number of things to make all this happen, from controlling 2 legs that are moving at opposite times, to moving your arms, to telling you when to bite, chew and swallow. I bet you can even tell me how the apple tastes and looks. Your body is taking in a lot of information, sending this information through nerves to your spinal cord, up to your brain to be figured out and then a response is sent back to the proper muscles. I can't imagine a computer handling this much information this quickly. Before all of this can happen though, information has to come into your body. You have five senses that bring in information. You have eyes, ears, a nose, a mouth and skin so you can see, hear, smell, taste and feel what's around you. Even when you are sleeping your brain is handling information. When you feel a feather, the information runs from your fingertips, through nerves, which are special, long cells that go to your spinal cord. Your spinal cord runs through the hollow center of your vertebrae and leads right up to your brain. Your nervous system is like a tiny super highway for messages.

Nervous System Lab #1: REACTION TIME - instructions

Material:

copy of lab sheet (1 page), pencil
standard (inch) ruler
colored pencil (for graphing)

Aloud: Your nervous system works incredibly fast to keep your hand from burning when you accidentally touch something that is too hot. Sometimes we are moving our hand away from danger even before we know it hurts. Your nervous system is also in charge of letting you taste good food as you are eating. All of this happens very quickly. Today we are going to do a lab to show how quickly a message can get from your eyes, along a system of nerves to your brain and then along another set of nerves all the way out to your hand. We are also going to see if you can increase the speed by practicing.

Procedure:

Lab Day:

1. Circle "Yes" or "No" for the Hypothesis part of the lab.
2. Have a helper hold the ruler about 1" above the catcher's hand with the zero end pointing down. Catcher should have their hand ready to pinch the ruler as it falls between his thumb and fingers (see diagram on lab sheet).
3. Helper drops the ruler with no warning. Catcher tries to catch the ruler as quickly as possible.
4. Note the point at which the catch was made (how many inches along the ruler). A smaller number means a quicker catch. Write the number on the chart.
5. Repeat above steps 4 more times.
6. To make a line graph, transfer data to the graph in the following manner: For catch #1 follow the horizontal line next to the label "CATCH #1" until you get to the inch mark where the ruler was caught. Place a dot there. Do this for each catch. Connect the dots to show how the reaction time changed. If the line goes right, reaction time got slower, if the line goes left, reaction time got faster (better).
7. Complete #4 with "got better," "stayed the same" or "got worse."

Conclusion / Discussion:

1. How did your reaction time change with practice. Can you train your nervous system to work faster?
2. Why do you think some people have quicker reactions to danger than others?

For More Lab Fun:

1. Test people of various ages. Does reaction time change with age? Compare and chart your results.
2. Test reaction time before and after exercise. Does exercise have an effect on reaction time?
3. Have the helper say, "Now," as they let go of the ruler. Does reaction time improve when you can "hear" the ruler drop?

NAME ______________________________ DATE ____________________

Nervous System Lab #1: REACTION TIME

HYPOTHESIS:

1. Do you think you can teach your body to react more quickly with practice?

YES NO

TEST:

2. On the chart below, write the inch mark where you caught the ruler.

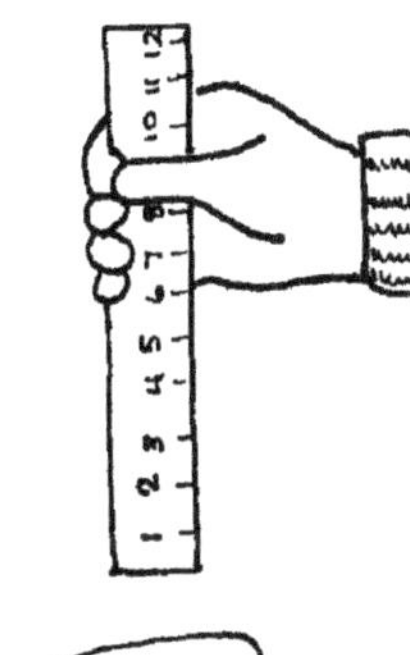

CATCH #1	CATCH #2	CATCH #3	CATCH #4	CATCH #5

RESULTS:

3. Use the information from your chart above to make a line graph. For each catch, go across the graph and make a dot at the inch mark where you caught the ruler. Draw a line to connect all five dots.

GRAPH OF MY REACTION TIMES

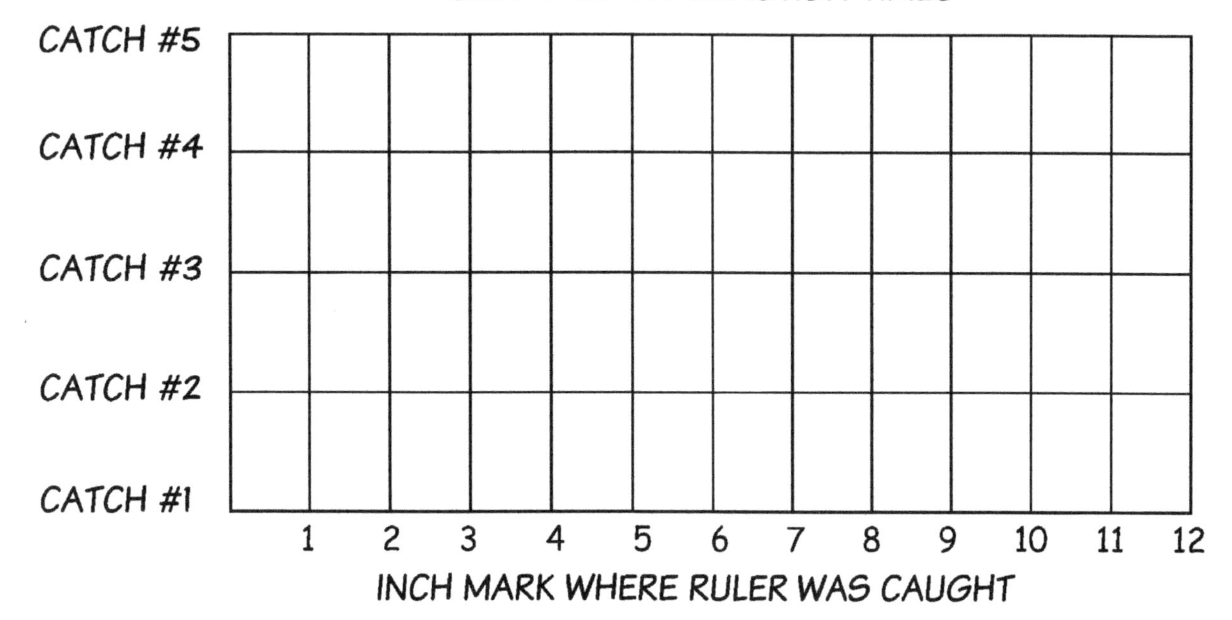

CONCLUSION:

4. My reaction time ______________________ with practice.

Nervous System Lab #2: I'M SENSIBLE - instructions

Material:

***This lab takes awhile to set up, but I have yet to meet a child (or adult!) that doesn't enjoy this lab.**

copy of lab sheets (2 pages), pencil

items to identify by taste, touch etc. This is a list of ideas. Use what you have and what your kids know.

touch - cotton swab, paper clips
smell - vinegar, garlic
taste - chocolate chip, banana slice
sight - orange, white vinegar
hearing - jingle bells, coins

4 small paper sacks to hide the touch and hearing items

3 small paper plates

2 spoons

3 small jars with lids- at least one MUST be see-through

4 paper towels to cover (hide) some of the items

dry-erase marker

Aloud: Step on a tack and you will jump away quickly. Smell cinnamon rolls and you will probably smile. A message has to get to your brain to tell you what is there and what to do about it. We talked about the long path the message has to take, but how does the message get into your body? You have five senses that bring information to your nerves. You have nerve endings all along your skin to tell you what you have touched, eyes to tell you what you have seen, a nose to tell you what you are smelling, ears to tell you what you are hearing and a tongue to tell you what you taste. All of these features bring information into your body, take that information to your brain and then tell your body what to do about it. And remember, it doesn't take very long. Today you are going to test your 5 senses to see how well they work to inform your brain of what is around you. Do your best but most importantly, follow directions and have fun!

Procedure:

Lab Day: **Lab Setup: FOR TEACHER/PARENT TO READ**

1. You will be setting up lab "stations" with different items to test. Try to keep them in line along a countertop or around a table.
2. On two paper sacks write "Touch." Place the 2 touch items inside and roll down the tops of the bags. These are stations 1 and 2.
3. Place the orange on a plate and about 1/4 c. of vinegar in a well sealed, see-through jar. Screw the lid down so no smell escapes. With the dry erase marker, write "Sight" on the plate and the jar. These are stations 3 and 4.
4. Put 1/4 cup more vinegar into another jar and some smashed, open garlic in another. Cover these with a paper towel and write "Smell only - No Peeking" on the towels. When it's time to test these you need to have the students close their eyes while you open the jars so they can smell what's inside. These are for stations 5 and 6.
5. On the other 2 sacks, write "Hearing." Place the hearing items inside these and roll the tops down for stations 7 and 8. Allow the children to shake these when it's time to test them.
6. On each of the last 2 plates place a spoon with the taste item in the spoon. You will need to spoon them into the children's mouths (with their eyes closed) for the test. Cover these with a paper towel also.
7. Go through the test with each child, assisting as needed to prevent "peeking."
8. Complete lab sheet.

CONTINUED ON BACK OF PAGE

Nervous System Lab #2: I'M SENSIBLE - instructions

Possible Answers:

Which sense is the most keen for each animal

Owl - hearing because they hunt at night, often in complete darkness

Vulture - smell because they eat rotting meat and must smell it from miles away

Bat - hearing because they also hunt at night and use their hearing to locate flying insects

Bald eagle - sight because they hunt during the day and use their sight to find prey.

Bloodhound - smell because they are specifically bred to find a scent and follow it. Their skin folds even cover their eyes when they put their heads down so they don't become distracted by what they see.

Conclusion / Discussion:

1. *How would it be to lose one of your senses? Try walking blindfolded (with a helper, for safety). Imagine eating without being able to taste your food. How would that be good or bad?*

For More Lab Fun:

1. *Kids always love to check out optical illusions to see how our senses often fool us.*
2. *Read a book about Helen Keller who lost both sight and hearing at a very young age. How different was her world from your own?*
3. *Watch one of your own pets. Try to discover by watching them which sense is the most important for them. Does your dog hear sounds that you cannot? What about your guinea pig? What seems to frighten him the most, a quick movement or loud noise?*

NAME ______________________________ DATE ____________________

Nervous System Lab #2: I'M SENSIBLE- pg. 1

For each station, use only the sense listed. Write what you think you have found and then write what other senses might have helped you to identify it.

A Station #	B Sense to Use	C Item (Name or Describe)	D Sense I would like to use
1	touch		
2	touch		
3	sight		
4	sight		
5	smell		
6	smell		
7	hearing		
8	hearing		
9	taste		
10	taste		

Nervous System Lab #2: I'M SENSIBLE- pg. 2

1. For each sense, this is how many I got right:

   touch ______ sight ______ smell ______ hearing ______ taste ______

2. It looks like my sense of ______________ is the most accurate.

3. This is how many times I wish I could have used each sense (column "D")

   touch ______ sight ______ smell ______ hearing______ taste ______

4. I wanted to use my sense of ______________ most often. This is the sense I trust the most.

5. (Circle one) The sense I trust the most is: **different than** **the same as** the sense that was the most accurate during this lab.

6. Every animal has a sense that is more keen than the others. This is their "dominant" sense. For each of the animals listed below, see if you can figure out which sense is the dominant sense.

Owl ______________________ because ______________________

Vulture ______________________ because ______________________

Bat______________________ because ______________________

Bald eagle______________________ because ______________________

Bloodhound ______________________ because ______________________

NAME ______________________________ DATE ____________________

For my notebook

GENES GUIDE YOUR GROWTH

Have you ever been told you were having a growth spurt? It's true that your body doesn't grow and change at a smooth rate. Growth and body changes happen in spurts. From the time you are born until the time you are very old your body will go through many amazing changes. Even before you are born the cells in your body carry the plans that determine when and how your body changes will occur. If you practice good healthy habits like eating and exercising right these plans can proceed at the proper rate and time. From the helpless baby you were at birth you have learned to walk, talk and dress yourself. At about 10 years old a body begins its journey to becoming its adult form. By around 18 - 20 years old most body systems are fully developed and upward growth stops. By continuing to make good, healthy choices you can keep your body fit and active throughout your adult life and well into old age. These plans in your cells are determined by your <u>genes</u> (jeenz). Genes are the special messages you get or <u>inherit</u> from your parents. Your genes help determine everything from how tall you are to how curly your hair is. Unless you have an identical twin, not one other person in the whole world has inherited the exact same body plan you have. From your one-of-a-kind fingerprints to your body's plans for your future growth, you are the only YOU there will ever be. Now that's a gift worth taking extra care of!

Genetics Lab #1: I'M THE ONLY ME! - instructions

Material:

copy of lab sheets (2 pages), pencil

Aloud: Your life started when one cell from your mom combined with one cell from your dad. These cells have tiny parts inside called genes. Your genes determine your hair color, sex, height and many other features. Except for identical twins, no two people share the exact same features; but you will share many features with your parents, because you inherit half of your genes from each one. Let's look at some traits to see what you have inherited.

Procedure:

Lab Day:

1. Examine the features listed for you and your parents (or siblings!) and fill in the chart.
   Explanation of some characteristics:
   - Cleft chin - chin with a deep "crease" down the middle
   - Free earlobes have a section at the bottom where they don't attach to the side of the head. Attached earlobes are attached to the head clear to the bottom of the earlobe.
   - Tongue rollers can curl their tongue into a "U" shape. This is not something a person can learn to do if they are born without the gene to do it!
   - Widow's peak is where the hairline forms a V down the forehead (think Eddie Munster).
   - Hairy knuckles refers to hair on the middle section of each finger, between the joints, not actually the knuckles.
2. For page 102, which of the traits do you share with your mother ONLY? Which do you share with your father ONLY? How many of the traits showed in neither parent and ended up showing in you?
3. Fill in the graph at the bottom. Color in how many traits you got from: Line 1: Mom ONLY Line 2: Dad ONLY or Line 3: from Mom and Dad- meaning they both show it (a dominant trait or they both do not show it (a recessive trait).

For More Lab Fun:

1. Make a chart to compare other relatives' features with your own. Which side of the family seems to have the strongest genes?
2. Start a family tree. Check old family photos to see if your ancestors had the same traits you have. How far back does that widow's peak go?
3. Visit a Labrador Retriever breeder. Can two black labs have yellow puppies? Can two yellows have black puppies? Which color seems to be weaker (recessive)?
4. Do you have a pen pal or friend from another country? Ask them to fill in the graph and chart the differences from one family-culture-geographic location to your own.

For Your Information:

1. Weak traits are called "recessive." Strong traits are called "dominant."
2. The traits given show a simple form of dominance. Some traits (not mentioned in this lab) are much more complex, some are linked to the sex chromosome and some traits blend.

NAME ______________________ DATE ______________

Genetics Lab #1: I'M THE ONLY ME! - pg. 1

All of the features below are determined by the genes you inherit from your mom and dad. Color in the blanks for each feature you have and compare.

	ME	MOM	DAD
CLEFT CHIN			
SMOOTH CHIN			
DIMPLES			
NO DIMPLES			
FREE EARLOBE			
ATTACHED EARLOBE			
TONGUE ROLLER			
NON-TONGUE ROLLER			
WIDOW'S PEAK			
ROUNDED HAIRLINE			
FRECKLES			
NO FRECKLES			
HAIRY KNUCKLES			
BALD KNUCKLES			

Genetics Lab #1: I'M THE ONLY ME! - pg. 2

I share these traits with my mom

_______________ _______________

_______________ _______________

_______________ _______________

____

I share these traits with my dad

_______________ _______________

_______________ _______________

_______________ _______________

____

Weak (recessive) traits (like dimples) can be carried by your parents but not show. If you get the same weak trait from both of your parents, even though it doesn't show in them, it will show in you.
These traits were secretly hiding in Mom and Dad and showed up in me!

_______________ _______________

_______________ _______________

_______________ _______________

I got this many traits from:

	1	2	3	4	5	6	7
Mom							
Dad							
Both							

I am exactly like.......ME!

Genetics Lab #2: MY OWN FINGERPRINTS - instructions

Material:

copy of lab sheet (1 page), pencil
1" wide transparent tape
3 extra people (to provide additional thumbprints)
hand lens (optional)

Aloud: Do you know what you have that nobody else in the world has? You have your own set of fingerprints that are unlike anybody else's in the whole world! With all the people in the world, it's hard to imagine that no two people would have the same fingerprints. Today you are going to analyze your fingerprints and see if you can match up a "mystery print" to the person it came from.

Procedure:

Lab Day:

1. On a piece of scratch paper, color in a heavy area of pencil. Make it dark enough that the pencil rubs off onto your finger. You may have to "reload" this area as you go along with the lab.
2. Starting with your left hand, rub your pinky in the pencil lead until it is coated. Have a helper tear off about a 1" piece of tape. Put the tape on the finger so that the print transfers to the tape. Remove from finger and tape it down in the first oval on your lab sheet.
3. Do the same for each finger on the left hand and then do the same for the right hand in the ovals on the line below.
4. Analyze each print. You may want to use a hand lens to help you. On the lines under each print, write whether that print is an arch, loop or whorl. (There are others, but these are the major ones. If your child has something different, choose the closest and call it good).
5. For #1, write how many of each print type you have.
6. Have 3 other people do a right thumbprint for you and place in the boxes below. Write the name of each print donor on the line below their print.
7. TEACHERS: Get an extra, mystery print from one of the donors- preferably the most unusual so it will be easier for the students to distinguish.
8. Compare the known prints with the mystery print. Write who you think the print belongs to on the line provided.

Conclusion / Discussion:

1. How can having individual prints help us?
2. How do we use fingerprints in our society? Could we do these things if people shared the same fingerprints?

For More Lab Fun:

1. Do fingerprint art with one of Ed Emberley's Print Drawing books.
2. Do fingerprints from both of your parents. Next do fingerprints of some friends or neighbors. Count how many arches, loops and whorls your parents have and then how many your friends have. Do your parents' prints have more similarities to yours or not? Why might your prints be more like your parents'?

NAME ______________________________ DATE ____________________

Genetics Lab #2: MY OWN FINGERPRINTS

These are the three basic types of fingerprints:

Arch Loop Whorl

My Left Hand

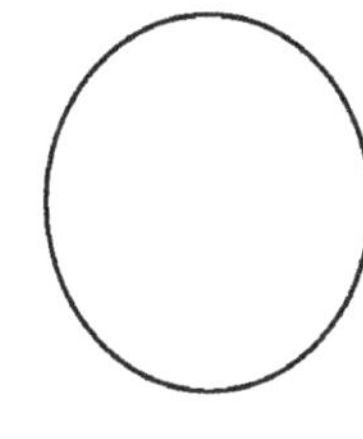
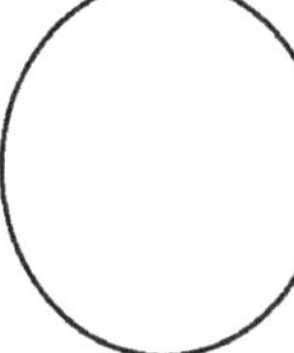

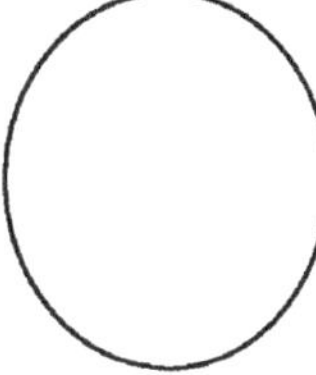

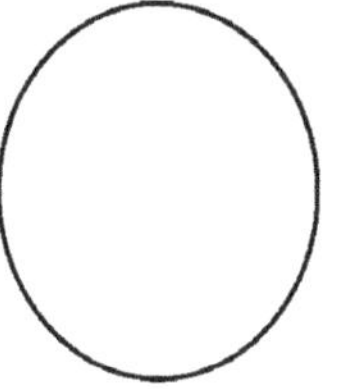
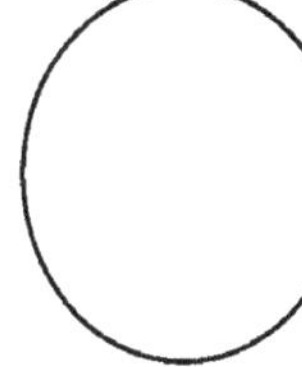

__________ __________ __________ __________ __________

My Right Hand

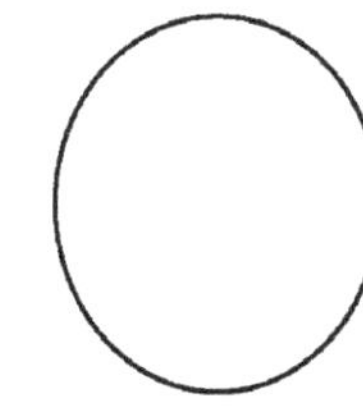
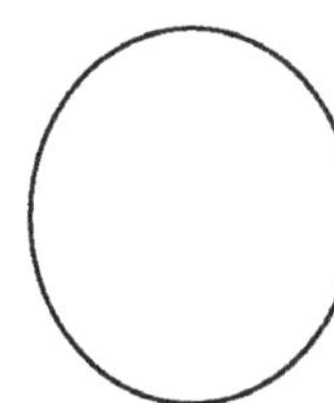

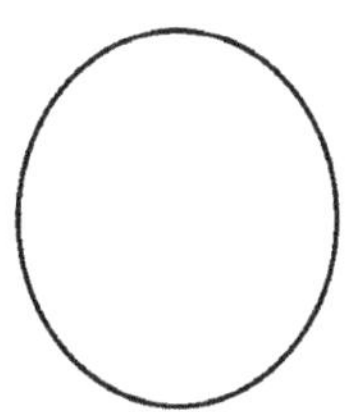
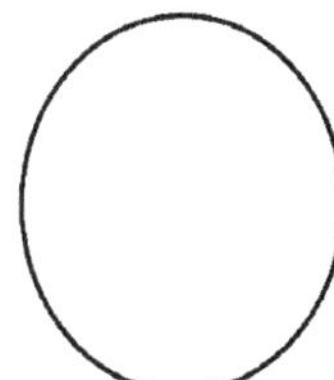

__________ __________ __________ __________ __________

1. I have _______ arches _______ loops _______ whorls.

This is the right thumbprint of:

Mystery Print!

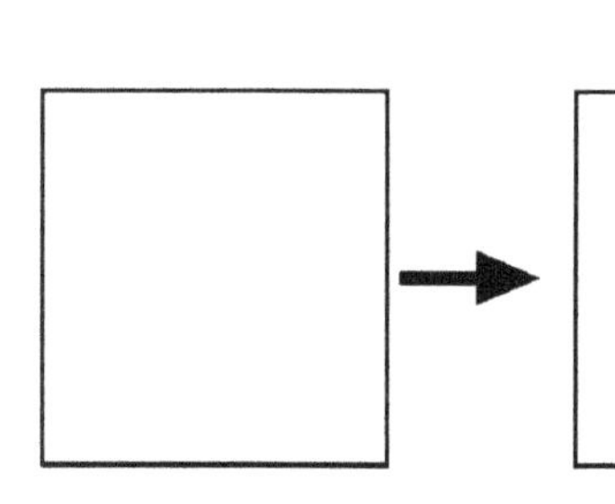

______________ (Name) ______________ (Name) ______________ (Name) ?

2. I think the mystery print belongs to ______________________________

NAME ______________________ DATE ______________

For my notebook

SIX KINGDOMS OF LIVING THINGS

You are about to learn about many different kinds of living things. You will also learn how to find, gather and investigate some common ones. To learn and share information about living things, scientists decided they needed to be able to group them somehow. To understand how important this is, imagine going to a store with no groupings. If there was no specific area for meat or bread you would have to search the whole store to find what you wanted. One loaf of bread might be next to a steak while the kind you want is under some broccoli. There are many more living things than there are items in a grocery store, so organizing them is vital. Most scientists agree that all living things can be split into six main groups called kingdoms. The two kingdoms we are going to learn about this year are the Plant Kingdom and the Animal Kingdom. Every kind of animal from a worm to an eagle is in the animal kingdom. Each kingdom can be split into smaller and smaller groups until you get to each species, (**spee**-sheez) like a dog, a bald eagle or a person. "Species" is a grouping, like kingdom but living things of the same species are very much alike. Bald eagles are one species. Every living thing has a scientific name which tells scientists what species it is. Tyrannosaurus rex is a scientific name that means "tyrant lizard." The scientific name of a person is Homo sapiens. That's you. The scientific name of an octopus is easy to remember. It's Octopus.

We will learn about many incredible living things, but first let's learn how scientists put all these different things into order.

Six Kingdoms Lab: CLASSIFYING CRITTERS - instructions

Material:

copy of lab sheets (3 pages), pencil
scissors
colored pencils (optional)

Aloud: Classifying animals is not as easy as it looks. Scientists do their best, but there are plenty of mistakes made and there will be plenty still left to discover when you are grown. Some animals look very much alike but are not the same species of animal. Other times animals that are the same species, don't look alike at all. Does a caterpillar look at all like the butterfly it will grow up to become? Male and female animals of the same species often look very different. Imagine a friend, exploring the planet Blobonia sends you Blobonian life forms in separate con-tainers and it is your job to pair up the male and female of each species. You will struggle to do your best but luckily, your friend has since had time to study the creatures in their natural habi-tat. Just in time, he sends along a classification key to help you in your quest. Use the key to learn how scientists divide animals into groups based on their characteristics.

Procedure:

Lab Day:

1. Carefully cut around each of the Blobonian life forms on page 1 of the lab. Make sure not to cut off the number near each one.
2. Go to lab page 2 and complete the hypothesis. Pair up the creatures as best you can and write your guesses down on the sheet.
3. Lo and behold the "Classification Key to Blobonian Life" arrives in your lab. Using one creature at a time, start on the left and follow the path that leads to the identity of each creature. When you get to where the creature belongs, write his/her number in the circle provided.
4. Complete the lab sheet.

Possible Answers:

#2. Animals pair as follows: 2 and 10; 8 and 9; 4 and 11; 1 and 6; 7 and 12; 3 and 5

#4. By studying nature scientists can find out what animals eat, how they live and which animals are really the same species. They can find out how animals live with each other, how they find food and how they avoid becoming food.

Conclusion / Discussion:

1. How important is it to observe living things in order to study them? People used to shoot things and study the bodies. What were they missing?
2. Without some system of grouping animals we wouldn't even have names like "bears" or "birds." Try to describe an animal like a beaver without using its name or any group names. How about describing a moose without using any group names (even names like horse or deer)?
3. How difficult would it be for people to share information about animals if we didn't have a classification system?

For More Lab Fun:

1. Imagine what the Blobonian world might be like. Color the Blobonians and them glue them to a decorated background.
2. Spread a big piece of butcher paper on the floor or table. Using the key as a guide, make your own classification key for several shoes. Start with all shoes together. Choose a characteris-tic to divide them in about half. Split each group further, writing down the characteristic you have chosen. Continue until you have each shoe in its own place. Bring in a "guest," give them a partner of one of the shoes and see if the key leads them to the correct shoe.

Six Kingdoms Lab: CLASSIFYING CRITTERS - Instructions

NAME ______________________ DATE ______________________

Six Kingdoms Lab: CLASSIFYING CRITTERS -pg. 1

NAME ______________________________ DATE ___________________

Six Kingdoms Lab: CLASSIFYING CRITTERS - pg. 2

HYPOTHESIS:

1. By looking at the specimens sent to me I would pair them up like this:

Pair #1: Number _____ and _____ Pair #2: Number _____ and _____

Pair #3: Number _____ and _____ Pair #4: Number _____ and _____

Pair #5: Number _____ and _____ Pair #6: Number _____ and _____

RESULTS:

2. In reality, the critters should be paired up like this:

Pair #1: Number _____ and _____ Pair #2: Number _____ and _____

Pair #3: Number _____ and _____ Pair #4: Number _____ and _____

Pair #5: Number _____ and _____ Pair #6: Number _____ and _____

CONCLUSION:

3. My pairing was **ALL RIGHT** **NOT ALL RIGHT**

4. These are some things a scientist can learn by observing living things in their own homes:

____

____

____

CLASSIFICATION KEY TO BLOBONIAN LIFE:

Use the key below to classify your critters. Once you know where each critter belongs, put its number in the circle provided. Each circle should have a different number in it. The two critters that end up together are the male and female of one species.

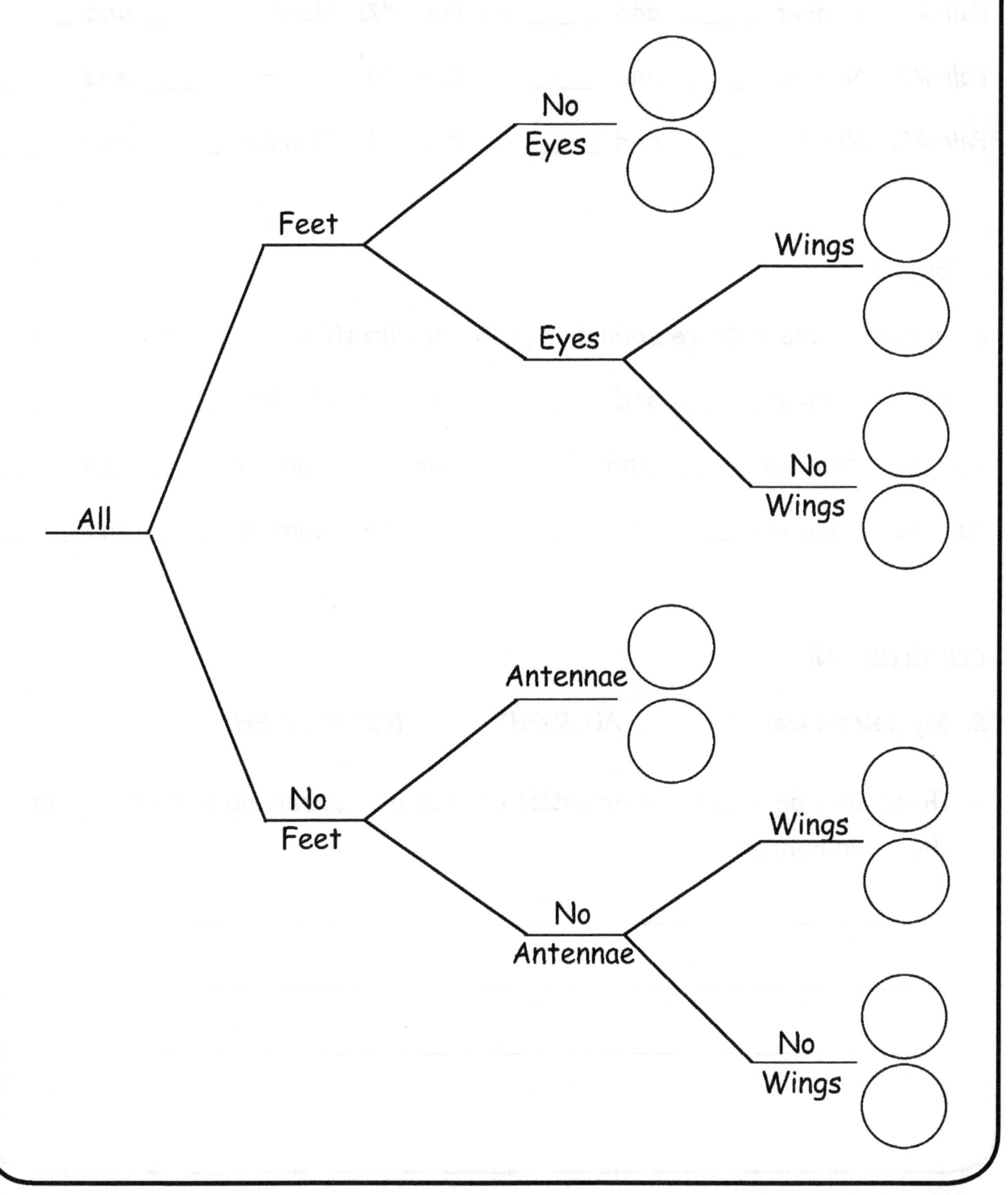

THE ANIMAL KINGDOM

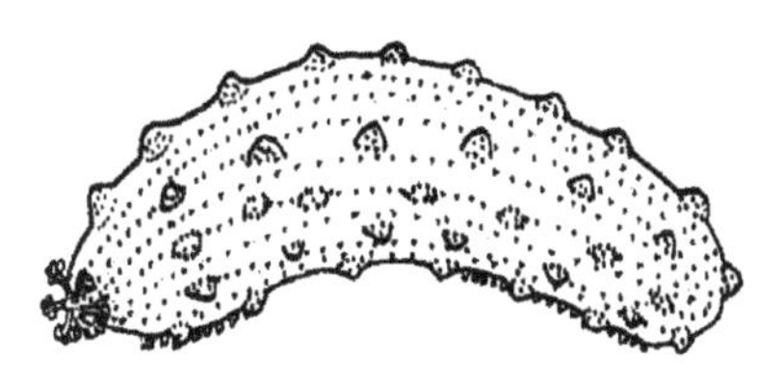

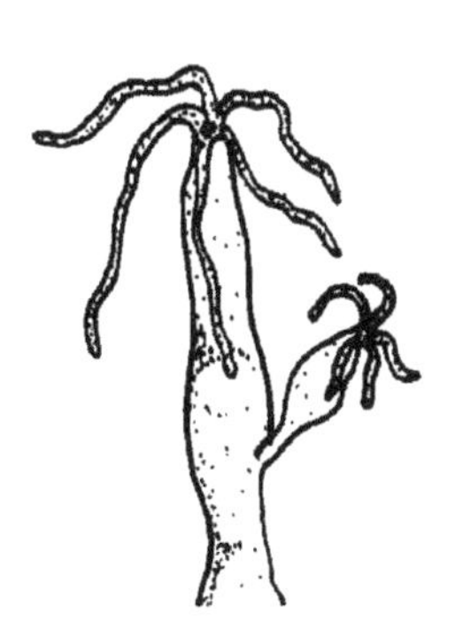

by

ANIMAL KINGDOM BOOK - instructions

As an ongoing project for the animal kingdom unit, your children will be assembling a booklet with a cover and a one page summary of each animal phylum (group). The book will be assembled ahead of time and then one new page will be finished each time your child gets to the end of studying another animal phylum. The assembly and cover will be done as a project today. You will find the summary pages and instructions for each additional page at the end of each phylum as it is studied.

Material:

- copy of "The Animal Kingdom" book
- cover page (page 115)
- 7 pieces of 12" X 12" cardstock - each a different color
- colored pencils
- scissors
- glue gel (like Elmer's school gel. It sticks better than glue stick, not as messy as white glue)
- stapler
- hole punch (if you want it to go into science binder)
- stickers, trim or any other decor your child wants to add to this page

Aloud: As we study the animal kingdom you will be putting together a book to show what you have learned. Today we will decorate the cover and assemble the book so it will be ready when you have information to add to it.

Procedure:

Lab Day:

1. Decide what order you want your cardstock pages in. The front and back covers will be uncut.
2. You will be cutting rectangles from the top right of 5 pages. Each cut will be 1 inch from the top. (Since rulers are usually marked in 1/16s of an inch I will give that measurement also so you just have to count little dots. For 3 4/16" you just measure in 3 and then count 4 little dots on your ruler past the 3). Hint: Mark where your cuts will be before cutting. Some 12" x 12" paper isn't exactly that.

   Page 1- cover: no cutting
   Page 2 cut: down 1" and cut in 9 1/2" (9 8/16")from the right and cut this piece off
   Page 3 cut: down 1" and 7 5/8" (7 10/16")from right and cut this piece off
   Page 4 cut: down 1" and 5 3/4" (5 12/16")from the right and cut this piece off
   Page 5 cut: down 1" and in 3 7/8" (3 14/16) and cut this piece off
   Page 6 cut: down 1" and in 1 15/16" and cut this piece off
3. Place all 5 cut pieces of cardstock together with the staggered notches at the top so that they all show. Set one uncut piece on top and one underneath and staple in 4 places about 1/2 inch from the left edge.
4. Color, decorate and cut around page 115. Add your name and glue onto the front cover of your Animal Kingdom Book. Each additional page is reserved for the summary of each animal phylum (group) as you reach that point. Add trim, stickers or whatever you have to make your cover irresistible.

For More Lab Fun:

1. Get pictures of yourself along the way, doing an experiment from each section. Add one picture to each summary page.

For Your Information:

As noted before, there are 6 kingdoms of living things. Each kingdom has a huge variety of living things and is therefore split further into phyla (singular-phylum) and each phylum split further. The entire living world is organized as follows, from most broad to most precise: Kingdom-Phylum-Class-Order-Family-Genus-Species. For years people have remembered this with some sort of word association. I learned "Kings Play Chess On Funny Green Squares." I have also heard "Kids Pick Chocolate Over Fancy Green Salads."

ANIMAL KINGDOM BOOK - Instructions

NAME ______________________________ DATE __________________

For my notebook

CNIDARIANS LOOK LIKE PLANTS

Have you ever seen *Finding Nemo* or another movie that shows an ocean floor full of corals and anemones (uh-**neh**-moe-neez) all flowing back and forth with the water? Not many years ago people thought all those beautiful things were plants. We have since decided that many of them are very simple animals. "Simple" means they don't have things like hearts, lungs or other organs but they are still animals. Why are they animals and not plants? Plants have green cell parts called chloroplasts, remember? With these they make their own food. Animals do not have chloroplasts and cannot make their own food. They must catch their food.

Most Cnidaria (nih-**der**-ee-uh) stick to the bottom of the ocean and seldom, if ever, move from place to place. Some, like sea jellies, are only stuck down part of their lives. The group cnidaria includes coral, sea jellies (often called jellyfish), sea anemones and one freshwater cnidaria, called a hydra. All cnidarians (animals in the group cnidaria) are just a hollow sack with a ring of stinging tentacles. The tentacles are to sting food so it can't get away. Coral is made of many tiny cnidarians all stacked together. Their tentacles wave in the water to catch food as it floats by.

So, all cnidaria are built like a hollow sack, surrounded by a group of stinging tentacles. They also have no complex organs. No wonder they are so often mistaken for plants.

Cnidaria Lab: SEA JELLIES CHANGE SHAPE - instructions

Material:

copy of lab sheet (1 page), pencil
colored pencils
4 stacking styrofoam cups
yarn- any color, about 3 yards
dry beans or seeds- one each of two different types
tape or glue

** You may choose to do this in one or two science classes. There are two parts to the activity. The first one is coloring in the life cycle chart. The other is building models of the life cycle to act out.*

Aloud: Most cnidaria are stuck down to the ocean floor with their tentacles waving in the water, waiting for food to pass by. Sea jellies are upside down cnidaria most of their lives but are stuck to the bottom part of their lives. When a cnidaria has its tentacles facing up it is called "polyp" (po-lip) shaped. When it's like an upside down cup with its tentacles facing down, it's called "medusa" shaped. Sea jellies change in their lives, just like butterflies do. Follow along on your life cycle picture. Jellies start as a section of a polyp. When a top section pops off, it flips over and becomes a medusa, or adult sea jelly. Sea jellies produce eggs and sperm, which meet in the water and grow into larvae which swim in the ocean. Eventually the larvae go to the ocean floor where they grow into other polyps and start to pop off their own little medusas!

Procedure:

Lab Day:

1. *Color in the sea jelly life cycle page, going over the names for the different stages of sea jelly life.*
2. *Cut the yarn into several 3" sections. Tape or glue them to the inside rim of 2 of the cups so that they hang down when the cup is turned upside down. They should hang something like a sea jellies tentacles.*
3. *Set the cups upright, letting the tentacles fall inside. If you glued the tentacles, wait for the glue to dry before going on to the next step.*
4. *Stack the cups carefully with the tentacles hidden inside. Set cups upright, open on the top.*
5. *Using the terms "polyp," "medusa," "egg," "sperm" and "larvae," help the children use the cups to act out the sea jelly life cycle. Cups stacked at the bottom of the ocean represent the polyp shape. Take out the first cup, slowly turning it over. As the tentacles fall out, it turns into an adult, medusa shaped sea jelly. If you would like, use the seeds to represent egg and sperm coming from the medusa and meeting in the water. Together they will form a larva which will eventually fall to the ocean floor and become a polyp and the whole cycle will start over.*

Conclusion / Discussion:

1. *Sea jellies are related to sea anemones and to corals, neither of which can swim through the ocean. How is it better, or worse to be able to swim freely through the ocean? (Swimming about makes it easier to find food but more likely to be eaten).*

For More Lab Fun:

1. *Read the greek myth about Medusa to find out what the medusa shape is named for.*
2. *Cut the cups in half and glue them to a piece of thick construction paper. Make a 3-D version of the life cycle, labeling all the stages.*

NAME ______________________ DATE ______________________

Cnidaria Lab: SEA JELLIES CHANGE SHAPE

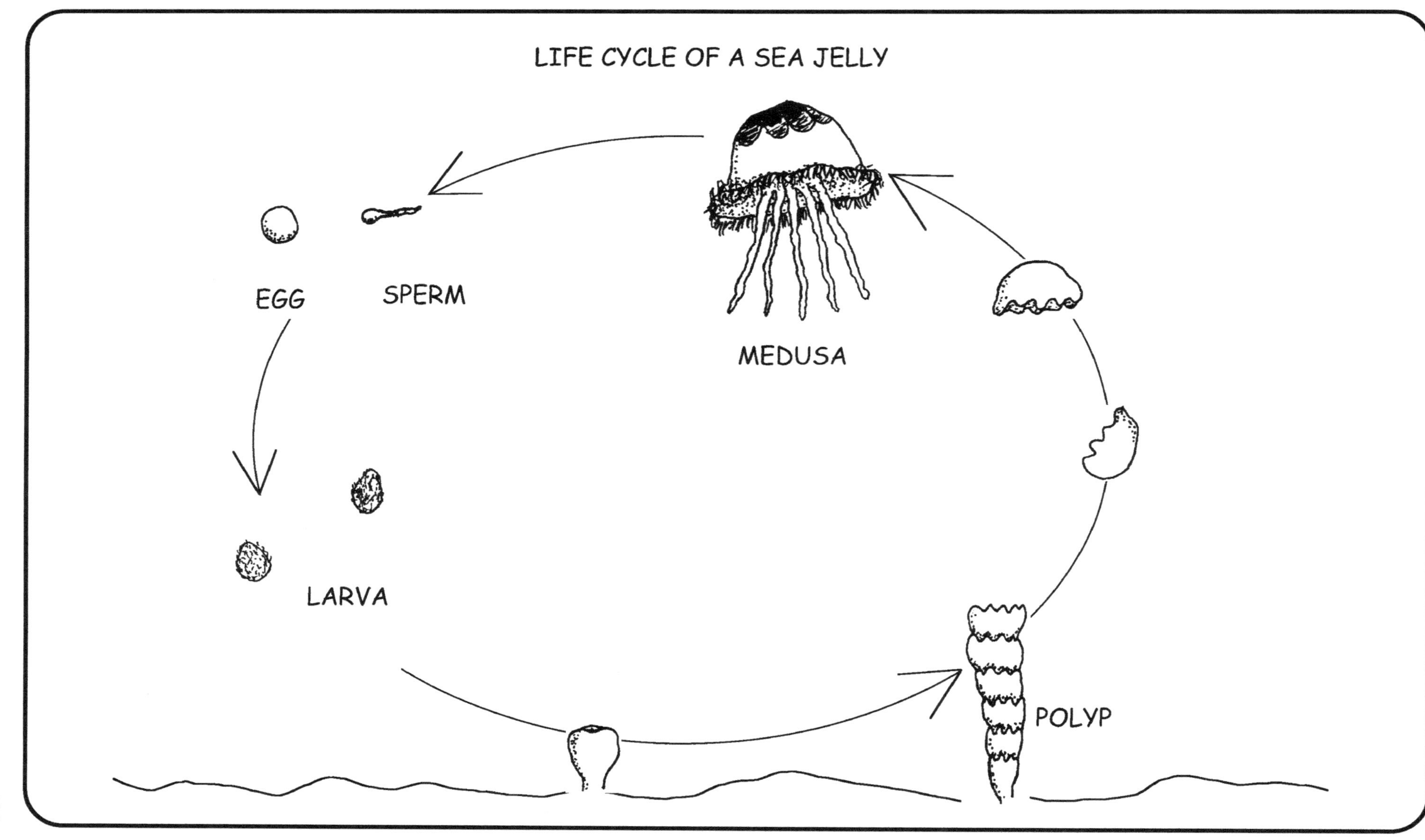

ANIMAL KINGDOM BOOK: CNIDARIA - instructions

Material:

copy of lab sheet (1 page), pencil or pen
Animal Kingdom Book (assembled)
scissors
glue gel (Elmer's School Glue Gel)
art supplies- markers, colored pencils, stickers, rubber stamps etc.
The Usborne Illustrated Encyclopedia of The Natural World (optional)

Aloud: Today we are going to do a summary of what we have learned about the phylum (fie-luhm) Cnidaria. Do your best, neatest work on this and you can create a wonderful summary of the animal kingdom as well as a beautiful book to treasure.

Procedure:

Lab Day:

1. Use the information from your science notebook to fill in sections of your lab paper. List the characteristics of cnidaria, examples of cnidaria and in the trivia section write the one thing about them that you thought was the most unusual or interesting. You may want to use the internet or *The Usborne Illustrated Encyclopedia of The Natural World* to find a neat fact about cnidaria.
2. Color the animals and the large cnidaria label on the page.
3. Page 1 (not the cover) in your Animal Kingdom Book is for the phylum cnidaria. Cut out the small cnidaria label (number 1) and glue it to page 1 at the top left where the tab is.
4. Cut out all remaining sections of your lab paper. Arrange them on page 1 of your book in whatever order you enjoy. Glue them in place.
5. Finish decorating this page as you wish.
6. Enjoy referring to your book to review what you have learned about the animal kingdom.

Possible Answers:

1. Characteristics of cnidaria:
   - look like plants
   - no complex organs
   - have stinging tentacles
   - built like a hollow sack
   - can be medusa or polyp shaped
   - most don't move from place to place
2. Examples of cnidaria:
   - coral
   - sea jelly
   - sea anemone
   - hydra (a freshwater cnidaria)

NAME ______________________ DATE ______________

ANIMAL KINGDOM BOOK - CNIDARIA

1. CNIDARIA

2. 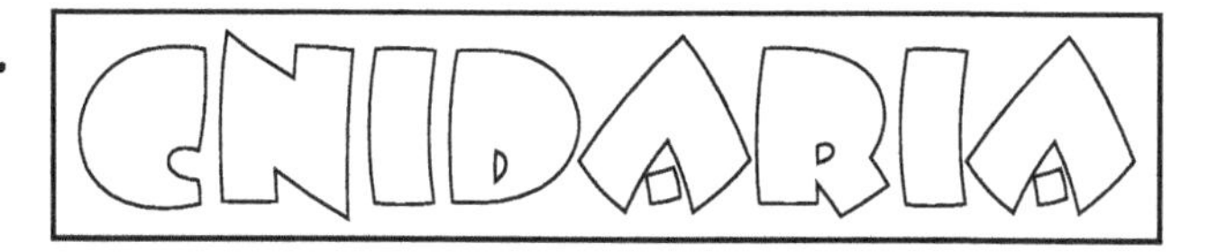

SEA ANEMONE

CHARACTERISTICS OF CNIDARIA

____

____

____

____

SEA JELLY

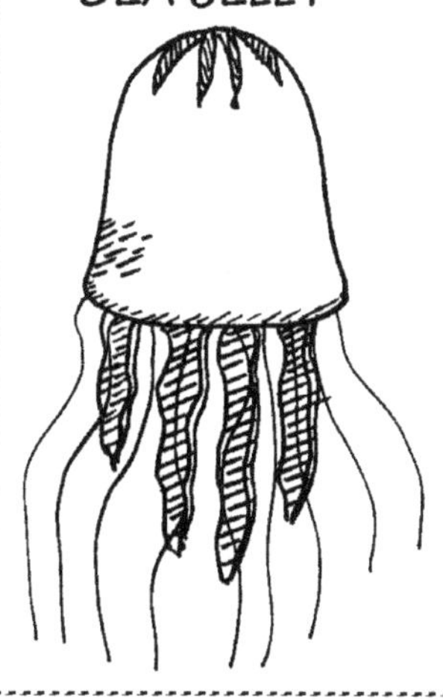

WOW!
CNIDARIA TRIVIA

____

____

____

____

____

____

EXAMPLES OF CNIDARIA

____

____

____

____

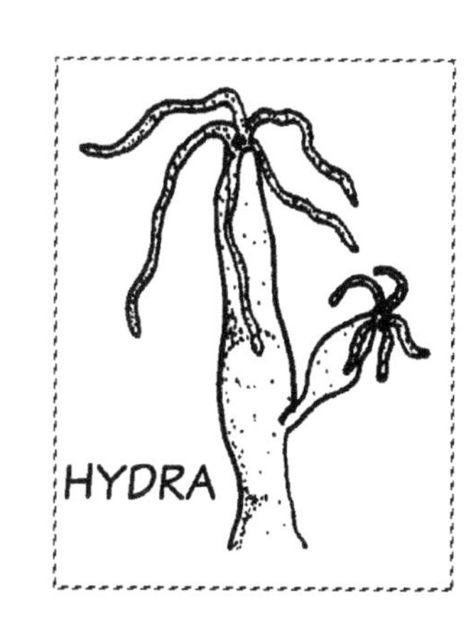

HYDRA

NAME ______________________________ DATE ____________________

For my notebook

WORMS ARE NOT INSECTS

Did you know that an earthworm can take our trash, like banana peels and newspaper, and turn it into healthy soil, perfect for growing a garden?

Earthworms are one of our best friends. Many worms are helpful to people but many can also be harmful. There are several diseases that people and other animals can get that are caused by different types of worms. How are worms alike? Worms are all shaped long like a string. They also all have a top and bottom side. Worms are much more complex than cnidaria inside. Do you remember that a cnidaria (coral or sea jelly) has no organs? Worms do. In fact an earthworm has five hearts!

Worms can be flat or smooth and round or divided into sections like a bunch of rings. Is an earthworm flat? Is it smooth or is its body in sections? The sections are easy to see, aren't they?

One thing worms don't have are legs. If you find a worm-like animals with legs, it's probably an insect.

So, worms do not have legs but they do have long bodies with a head and tail, a top and bottom and pretty complex organs inside. Some can be harmful and some actually help keep us alive. Next time you see an earthworm, thank him for his help in the garden.

Worm Lab #1: THE GREAT WORM HUNT - instructions

Material:

copy of lab sheets (2 pages), pencil
moist soil - NOT SOAKED
plastic jar or other waterproof container with lid
leaves, lawn clipping, oatmeal, lettuce . . . (any or all of these for food)
flashlight (optional)
red cellophane to fit over front of flashlight (optional)
pitchfork (causes less damage to worms than a shovel does)

Aloud: We are going to hunt for earthworms. They seem easy to find when we aren't looking. Can you think of some places you have found them before? Let's guess where they might like to live and then go out and test our guesses. We will be saving our worms for the next labs.

Procedure:

Lab Day:

1. Prepare worm home with moist soil. Add some leaves, a little lettuce or lawn clippings. Remember – worms can drown because they breathe through their skin. This is why you see so many earthworms on the pavement after a rain. They have crawled out of the soil to avoid drowning. With this in mind, make sure the soil is moist but not soaked.
2. Fill out page one of lab sheets.
3. Allow children to direct the worm hunt. Have them try in places they have guessed the worms might hide. You can even try two hunts- one in the light and one in the dark. If you go out in the dark, try putting red cellophane over the flashlight. Earthworms aren't bothered by red light but will hide from white light. If you are unsuccessful (for some odd reason) try watering your lawn right at bed time and see what comes out in the dark. Keep track of how many worms you found in different locations. Word of caution – worms have bristles that hold them in the ground when being pulled. Be careful not to let kids pull them in half trying to extract them from the soil. Instead, pat the dirt off of them after digging them up.
4. Fill in page #2 of your lab sheet. Choose one large worm to study and draw. Make sure it is one with a smooth "cuff" on it.

Aloud: Feel the bottom of your earthworm. Can you feel the bristles? These help the worm move through the soil and help him hold onto the soil when a bird or other predator is trying to pull him out. Find his mouth. It is closest to the end with the smooth cuff, or clitellum (klih-tell-um). Worms are both male and female. They use the clitellum for laying eggs.

5. Label the worm drawing.

Possible Answers: (Page 2)

6. 0 antennae 7. 0 legs 8. 0 wings 9. 0 eyes
10. up to 100 body segments, depending on the species of earthworm

Conclusion / Discussion:

- Determine and discuss where the most worms were found. Use this information to set up your worms' home.
- How might their bristles help them when a bird grabs onto them and tries to pull them out of the soil?
- Don't put your worms back outside yet. You will use them for the following worm labs as well.

NAME ______________________________ DATE ___________________

Worm Lab #1: THE GREAT WORM HUNT - pg. 1

I predict I will find the most worms:

1. ☐ after dark ☐ during the day

In soil that is:

2. ☐ dry ☐ moist ☐ soaked

3. ☐ bare ☐ grassy areas ☐ under rocks or logs

4. ☐ above the ground ☐ underground

5. I think earthworms look like this:

I think earthworms have: (Write the number next to each feature)

6. ________ antennae

7. ________ legs

8. ________ wings

9. ________ eyes

10. ________ body segments

Earthworm Lab #1: THE GREAT WORM HUNT - pg. 2

I really did find the most worms:

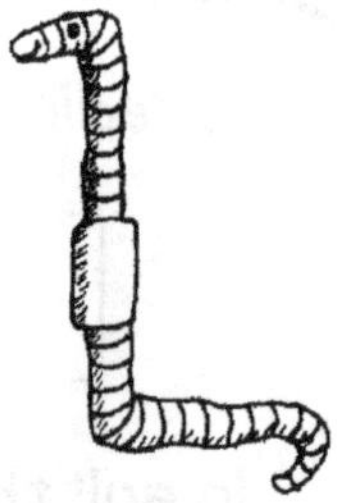

1. ☐ after dark ☐ during the day

In soil that was:

2. ☐ dry ☐ moist ☐ soaked
3. ☐ bare ☐ grassy areas ☐ under rocks or logs
4. ☐ above the ground ☐ underground

5. My favorite earthworm looks like this:

On your earthworm label the following: mouth, segments, bristles, clitellum
Use the diagram below to help you

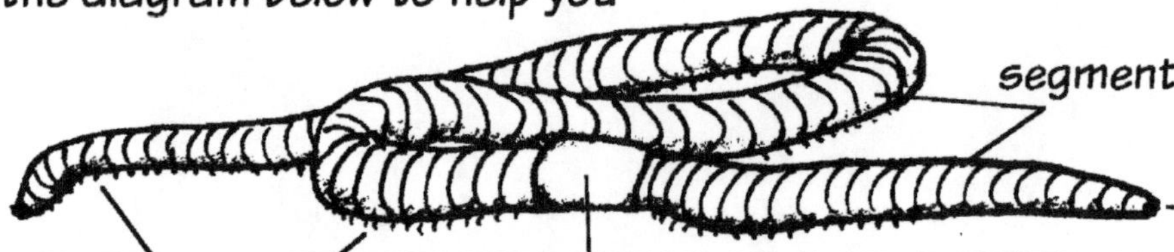

segments
mouth
bristles
clitellum

My earthworms have: (Write the number next to each feature)

6. ________ antennae
7. ________ legs
8. ________ wings
9. ________ eyes
10. ________ body segments (my best guess!)

Worm Lab #2: EARTHWORM COMPOSTING - instructions

Material:

- copy of lab sheets (2 pages), pencil
- 5 or 6 earthworms (you can buy them at a bait shop or dig them up yourself)
- tall see-through jar (mayonnaise size or bigger) with holes punched in lid
- large paper grocery sack big enough to cover jar
- tape
- sand
- soil
- oatmeal
- timer or watch with second hand
- mister with water

Aloud: Who works all day and night, requires no pay, never complains, recycles the garbage and helps to grow healthy food for you and me? Earthworms. Earthworms can take newspaper, coffee grounds, rotten lettuce, eggshells and many other things we consider trash and they can make good, healthy soil. They also tunnel through the soil, breaking it up and moving air and water into it to make it even more healthy. Healthy soil grows healthy food. Today you are going to set up an experiment to see how earthworms mix soil to make a garden healthy. Please be gentle with your worms. They are worth more than gold!

Procedure:

Lab Day:

1. In your jar layer moist sand, then oatmeal then moist soil. Repeat this layering. Make sure the layers are moist but not wet. Earthworms will drown in wet soil. You should have two layers of each item and leave a space of an inch or more at the top.
2. Use a marker, to draw a mark on the jar at the top of each layer. Measure the layers and record on the lab paper.
3. Add worms and watch what they do. As soon as you put them in the jar, start the timer. Note how long it takes them to burrow into the soil. Fill in the lab sheet as indicated.
4. Place lid on jar. Cover with grocery sack so no light enters. Place in a dark, cool place. Keep soil moist throughout experiment.
5. Check and mist jar daily. When layers are all mixed together, do lab page #2.
6. Save your worms for lab #3.

Conclusion / Discussion:

- Discuss results of lab. Explain how soil must be mixed and aerated in order to be healthy. This can be very time consuming to do by hand but earthworms do it for us for free!

For More Lab Fun:

1. Start a compost pile and add your worms. Add leaves, grass clippings, food scraps, old shredded newspapers, shredded carrots or crushed eggshells. Use your compost to help grow your own garden.
2. Keep your worms in the jar. Every week or so add different foods to compare how long it takes the worms to mix different items. Try any of the foods mentioned above. If you keep them for a while, check for eggs and tiny baby worms.

NAME ____________________ DATE ______________

Worm Lab #2: EARTHWORM COMPOSTING - pg. 1

The world's largest earthworms can grow to over 20' in length!

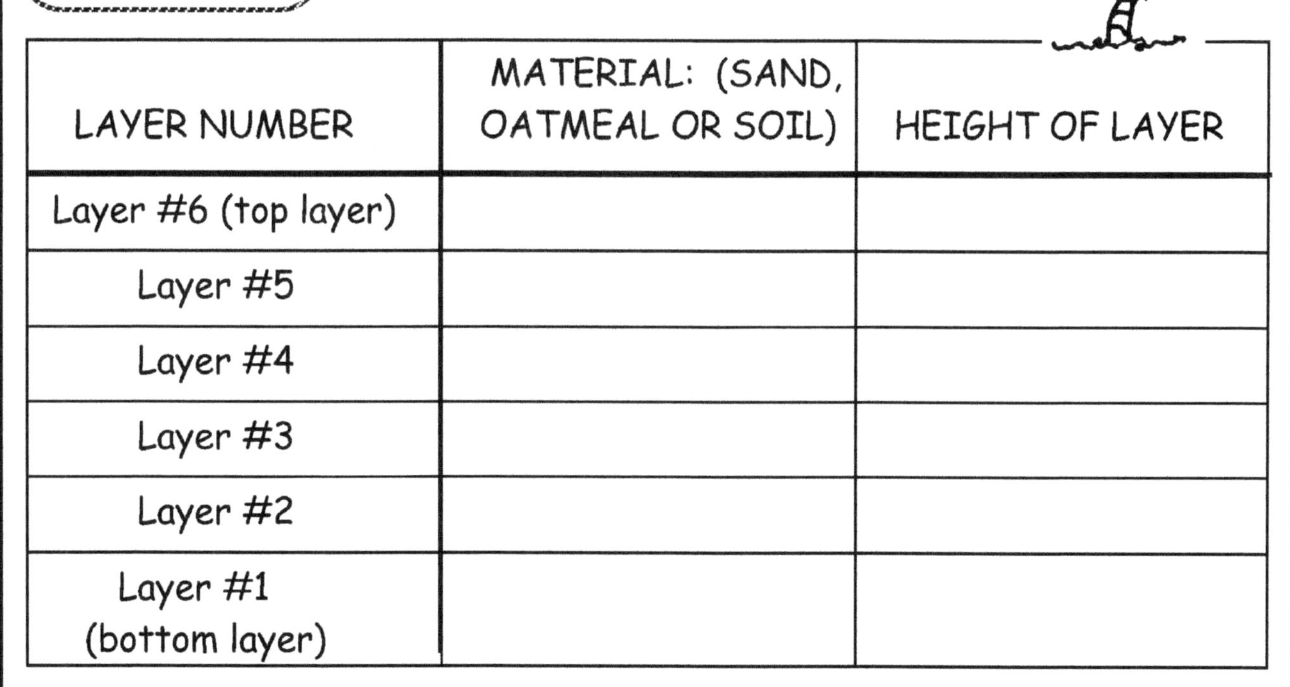

LAYER NUMBER	MATERIAL: (SAND, OATMEAL OR SOIL)	HEIGHT OF LAYER
Layer #6 (top layer)		
Layer #5		
Layer #4		
Layer #3		
Layer #2		
Layer #1 (bottom layer)		

1. Describe the layers in your jar. Is it easy to tell the layers apart?

2. What did the earthworms do when you put them into the jar?

3. If they dug into the soil? How long did it take them?

Worm Lab #2: EARTHWORM COMPOSTING - pg. 2

AFTER WORMS: My compost jar looks like this:

Compare your before and after pictures

4. *Describe the layers in your jar. What has changed since you set it up?*

5. *What have the earthworms done ?*

6. *How long did it take for the worms to mix the soil completely?*

7. *How does this help farmers?*

Worm Lab #3: EARTHWORMS AREN'T SENSELESS - instr.

Material:

copy of lab sheets (2 pages), pencil
earthworm
tray with sides- a 9" X 9" baking pan, food storage container, etc.
2 paper towels
aluminum foil- enough to cover half of the top of the tray
flashlight
source of hot and cold water
small amount of vinegar
piece of rough sandpaper
piece of smooth binder of copy paper

Aloud: We use our eyes and ears to learn about the world around us. To stay alive, an earthworm needs to know what is happening in its environment too, but it has no eyes or ears. Every segment of an earthworm has special cells that can detect changes. In the lab today, you will test a worm to see what kinds of things these cells can detect.

Procedure:

Lab Day:

1. Place a moist paper towel on the bottom of the pan so your earthworm can breathe.
2. Fill out the predictions for the lab.

TEST I

3. Cover half the pan (over the top) with foil. Hold the flashlight so it shines down onto the uncovered half. Place the worm in half light and half shade. Record in which direction the worm goes.

TEST II

4. Place a moist paper towel on half of the pan bottom. Place a dry paper towel on the other half making sure they meet in the middle of the pan. Place the worm half across the wet and half across the dry towels. Record his preference.

TEST III

5. Dip one paper towel in hot water, the other in cold. Place each towel on half of the pan. Again, place your worm across both sides and record his preference.

TEST IV

6. Cover half the pan bottom with sand paper and the other half with smooth binder or copy paper. Test and record results.

TEST V

7. Dip a strip of paper into the vinegar. Hold this near the worm WITHOUT TOUCHING HIM! Record his reaction.

TEST VI

8. Have children make up their own test.
9. Return worms to the soil or to your garden

Conclusion / Discussion:

1. In theory your worms should move towards dark, moist, cold, smooth and away from vinegar. You may get different results but this is the usual.
2. Discuss with children how this is most like the environment in which they live. They would dry out in sunny, dry and hot places. As for the vinegar, remarkably, earthworms have a great sense of smell.

For More Lab Fun:

1. Test your worms response to different colors of light by placing colored cellophane over the flashlight. (Earthworms move away from all light except red.)

NAME ______________________ DATE ______________________

Worm Lab #3: EARTHWORMS AREN'T SENSELESS- pg. 1

TEST I: REACTION TO LIGHT

I predict my worm will like ◯ light ◯ dark

because ______________________

Actually, my worm prefers ◯ light ◯ dark

Where worms live it is ______________________

TEST II: REACTION TO MOISTURE

I predict my worm will like ◯ wet ◯ dry

because ______________________

Actually, my worm prefers ◯ wet ◯ dry

Worms prefer soil that is ______________________

TEST III: REACTION TO TEMPERATURE

I predict my worm will like ◯ hot ◯ cold

because ______________________

Actually, my worm prefers ◯ hot ◯ cold

Worms prefer soil that is ______________________

TEST IV: REACTION TO TEXTURE

I predict my worm will like ◯ smooth ◯ rough

because ______________________

Actually, my worm prefers ◯ smooth ◯ rough

Worms prefer soil that is ______________________

Worm Lab #3: EARTHWORMS AREN'T SENSELESS- pg. 2

TEST V: REACTION TO CHEMICALS

I predict my worm will ◯ likes vinegar ◯ dislikes vinegar

because ______________________________

Actually, my worm ◯ likes vinegar ◯ dislikes vinegar

Where worms live it is ______________________________

TEST VI: OF MY OWN DESIGN! ______________________________

I predict my worm will ◯ ◯

because ______________________________

Actually, my worm ◯ ◯

Worms ______________________________

This is a picture of my own test

According to my test results, the perfect environment for a worm would be:

____

____

____

____

ANIMAL KINGDOM BOOK: WORMS - instructions

Material:

copy of lab sheet (1 page), pencil or pen
Animal Kingdom Book (assembled)
scissors
glue gel (Elmer's School Glue Gel)
art supplies- markers, colored pencils, stickers, rubber stamps etc.
The Usborne Illustrated Encyclopedia of The Natural World (optional)

Aloud: Today we are going to do a summary of what we have learned about worms. You will be adding another page to your Animal Kingdom book. Remember to do your best work on this as it should last and help you for years to come.

Procedure:

Lab Day:

1. Use the information from your science notebook to fill in sections of your lab paper. List the characteristics of worms, examples of worms and in the trivia section write the one thing about them that you thought was the most unusual or interesting. You may want to use the internet or *The Usborne Illustrated Encyclopedia of The Natural World* to find a really neat fact.
2. Color the animals and the large worm label on the page.
3. Page 2 in your Animal Kingdom Book is for your worm summary. Cut out the small worm label (number 1). Have your book open to page 1 (cnidaria). The worm label will go on the small tab for page 2 that shows at the top and to the right of the cnidaria label. Glue your worm label in place.
4. Turn to page 2. Cut out all remaining sections of your lab paper. Arrange them on page 2 of your book in whatever order you enjoy. Glue them in place.
5. Finish decorating this page as you wish.
6. Enjoy referring to your book to review what you have learned about the animal kingdom.

Possible Answers:

1. Characteristics of worms:
   - shaped like a string
   - more complex organs
   - have a top and bottom
   - no legs
   - have a head and tail
2. Examples of worms:
   - earthworm (segmented worm)
   - planaria (flatworm)
   - leech (segmented worm)
   - hookworm (roundworm)
   - tapeworm (flatworm)

NAME ______________________________ DATE ____________________

ANIMAL KINGDOM BOOK- WORMS

1. WORMS

2. 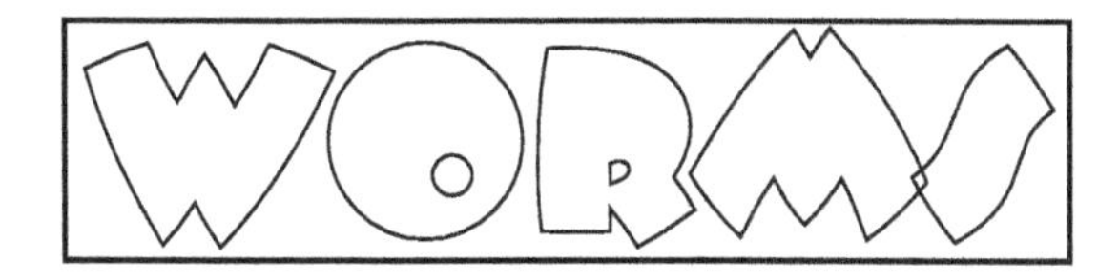

CHARACTERISTICS OF WORMS

____

____

____

____

EARTHWORM

AWESOME WORM TRIVIA

____

____

____

____

____

____

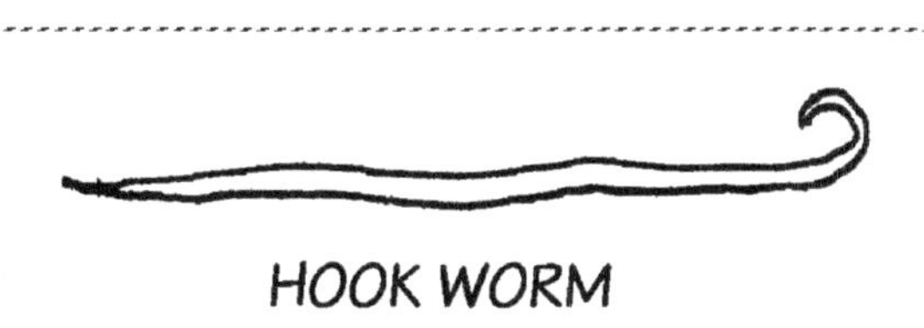

HOOK WORM

EXAMPLES OF WORMS

____

____

____

____

PLANARIA

NAME ______________________ DATE ______________

For my notebook

MOLLUSKS ARE SQUISHY

I wonder if you have ever eaten a mollusk (mol-usk). Well, people all over the world eat almost every type of mollusk. So, what are mollusks? Mollusks are clams, oysters, octopi, squids, snails and others. Have you eaten any of these? It's also fun to have races with them and just watch them move along "at a snail's pace."

It's hard to imagine how animals as different as a snail, a clam and an octopus ended up in the same group but let's look more closely at them. Remember we talked about how simple a cnidaria was, with no real organs inside? Mollusks have pretty complex organs. Just like you, they have a digestive system, circulatory system and nervous system. All mollusks have a soft, squishy body. Clams and snails hide theirs inside a hard shell but their bodies are still soft.

The third reason mollusks are grouped together is because of their hard shell. Clams and oysters have 2 shells and the different land and water snails have one shell. Most animals from the third group have shells inside where you can't see them but they all have 8 or more tentacles. The octopus and squid are in this group.

So, as different as they look, all mollusks have complex organ systems and a soft body and almost all have a hard shell either inside or out.

Mollusk Lab #1: MOLLUSK? WHO ME? - instructions

Material:

copy of lab sheets (2 pages), pencil

a few garden snails to observe (see the Critter Care Sheet pg. 11 for finding and housing snails)

see-through jar

hand lens

Aloud: Can a snail really be related to an octopus or to a clam? That doesn't seem possible. Snails and slugs live on land but the other Mollusks live in water. Today we are going to study a garden snail and see how it compares to other mollusks and to animals that are not mollusks. Be gentle with your snail as their shells can be very delicate.

Procedure:

Lab Day:

1. *Take time to gently handle and play with your snail.*
2. *Use your snail and your Mollusk Notebook Page to fill in the chart on lab page 1.*
3. *Compare each animal with your snail. For each animal, put a tally mark for every way it is like your snail. If both have a characteristic or if both do not have it, give it a tally mark. Don't put a tally only if one animal has the characteristic and the other doesn't. For example, if your snail and the earthworm both have a shell, tally it. If your snail and the earthworm both do NOT have a shell, put a tally mark, because they are alike in that way. If the snail has a shell and the earthworm doesn't, don't tally it because they are different. It is easiest to go through one animal at a time, comparing each characteristic to the snail as you go.*
4. *Add up the tally marks and record for each animal on #3.*
5. *Whichever animal has the most tallies is the most similar to a snail. Write this down on #4.*
6. *Let your snail crawl up the inside of a jar. Look at the underside of his foot through the glass. Watch the wave-like motion of the foot muscle. Also from underneath the snail, use your hand lens to look at the mouth. You may actually be able to see inside at the tongue. It is covered with thousands of tiny hooks for scraping its food.*
7. *Take a close look at the slime trail your snail leaves. This helps your snail glide along more smoothly and, because the slime hardens as he goes, it also prevents him from falling backwards when he is climbing.*

Possible Answers: *TURN TO BACK OF PAGE FOR ANSWERS*

Conclusion / Discussion

1. *Often it is hard to tell why scientists have grouped animals the way they have but usually with closer inspection, the reasons become more clear.*

For More Lab Fun:

1. *Snails are easy to keep. If you have a few snails in your new snail home, watch for eggs. They will often show up under your piece of log. Look at one closely with a hand lens. Just remember they need food and moist but not wet, soil.*
2. *Do snail slime art. Let your snail crawl around on a piece of black construction paper. Put your snail away. Run a thin strip of glitter glue where your snail crawled. Voila! Snail art.*

Possible Answers:

Page 1:

	SNAIL	CLAM	OYSTER	SEA JELLY	EARTH WORM	BEETLE
Soft body	X	X	X	X	X	
Hard shell- 1-2 parts	X	X	X			
Complex organs	X	X	X		X	X
Stinging tentacles				X		
String shaped					X	
Jointed legs						X

2. Use tally marks to compare your snail with the other animals listed. If the snail and the other animal are alike (either both have or both do not have the characteristic) put a tally mark.

Clam 𝍸 | Oyster 𝍸 | Sea Jelly ||| Worm |||| Beetle |||

3. Add up the tally marks for each animal. Write the total for each next to the name of that animal below. Compare your totals to find out which animal your snail is most like.

Clam 6 Oyster 6 Sea Jelly 3 Worm 4 Beetle 3

Page 2:

#4 clams and oysters

#5 clams and oysters

mollusk group

#6 mollusk

all mollusk have soft bodies

NAME ______________________________ DATE ________________

Mollusk Lab #1: MOLLUSK? WHO ME?- pg. 1

1. Fill in the chart below for your snail. Place an X in each box if your snail has that characteristic. The other animals are done for you. Look at your snail and your Mollusk Notebook Page to find the answers.

	SNAIL	CLAM	OYSTER	SEA JELLY	EARTH WORM	BEETLE
Soft body		X	X	X	X	
Hard shell- 1-2 parts		X	X			
Complex organs		X	X		X	X
Stinging tentacles				X		
String shaped					X	
Jointed legs						X

2. Use tally marks to compare your snail with the other animals listed. If the snail and the other animal are alike (either both have or both do not have the characteristic) put a tally mark.

Clam______ Oyster______ Sea Jelly______ Worm______ Beetle______

3. Add up the tally marks for each animal. Write the total for each next to the name of that animal below. Compare your totals to find out which animal your snail is most like.

Clam______ Oyster______ Sea Jelly ______ Worm______ Beetle_____

Mollusk Lab #1: MOLLUSK? WHO ME?- pg. 2

4. According to my tally, snails are most like______________________________

5. Clams and oysters are in the mollusk group.
   Sea jellies are in the cnidaria group.
   Earthworms are in the worm group.
   Beetles are in the arthropod group. (We'll learn more about arthropods later).

   If snails are most like ____________________________________ then

   they must be in the ______________________________ group.

6. In Latin, "mollis" means soft. This word sounds most like the group name

   ________________________________. I think this group is called "mollis"

   because ____.

7. According to my study, snails (circle one): should should not
   be in the mollusk group with clams, oysters and squids.

8. Other things I got to see on my snail:

   - ☐ Wave-like muscles on the bottom of his/her body (foot)
   - ☐ His/her mouth
   - ☐ His/her slime trail

Mollusk Lab #2: SNAIL ANATOMY - instructions

Material:

- *copy of lab sheets (2 pages), pencil*
- *garden snail to observe (saved from Mollusk Lab #1)*
- *centimeter ruler*
- *gram scale or triple beam balance*
- *hand lens*
- *various nontoxic plant items to offer your snail- lettuce, broccoli, cabbage, ivy, cucumber etc.*

Aloud: Garden snails and slugs are land mollusks. We're going to measure and watch ours today and learn a little more about how they live. Snails have tentacles that feel along and tell them about the environment. Gently touch a tentacle and see how your snail reacts. The top two tentacles contain the snail's eyes. You can see the nerve that runs from the eye down the tentacle. This carries the message to the snail's brain about what the snail is seeing. The bottom two tentacles "smell" the environment. Snails move along on a soft "foot" that is a big part of their bodies. Are you wondering if your snail is a boy or girl? Well, you're right. It is a boy AND a girl. Snails are both!

Procedure:

Lab Day:

1. *Do the lab sheet, page 1 as indicated.*
2. *Do your best to measure the parts of the snail as indicated on #4, page 2.*
3. *Encourage your budding artists to do the best they can on the diagram. Include all parts that they have observed.*

Possible Answers:

#1. eyes = seeing shell = protecting the soft body foot = movement tentacles = smelling mouth = eating

#2.

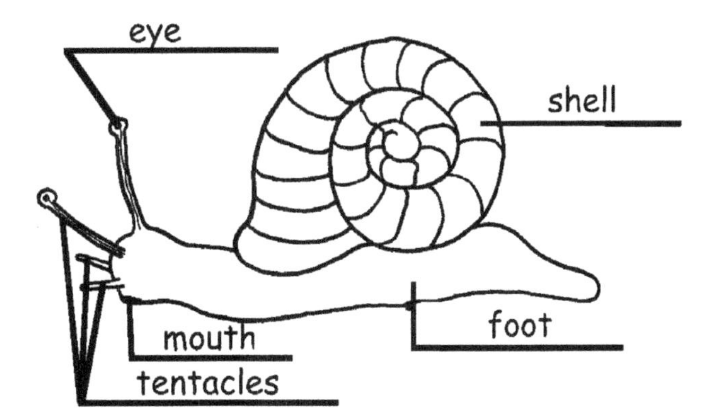

For More Lab Fun:

1. Have a snail meal. Eat only food you have grated. Try cheese, carrots, cucumber, cole slaw and whatever else comes to mind. Finish off with cookies that contain grated lemon peel (zest).
2. Hold snail races. Time your snails in various events such as the ramp climb. Measure their distance in events such as the 1 minute "dash."

Mollusk Lab #2: SNAIL ANATOMY - Instructions

Materials:

[illegible]

[illegible]

[illegible]

[illegible] or trip [illegible] the ince

hand [illegible]

various nontoxic plant items to offer your snail—lettuce, broccoli, cabbage, [illegible] etc.

[illegible] about Garden snails and slugs [illegible] land mollusks. We're going to measure and watch [illegible] ly and learn a little more about how they live. Snails have tentacles that feel along and tell them about the environment. Gently touch a tentacle and see how your snail reacts. The [illegible] two tentacles [illegible] the snail's eyes. You can see the nerve that runs from the eye down the tentacle. This carries the message to the snail's brain about what the snail is seeing. The bottom two tentacles "smell" the environment. Snails move along on a soft "foot" that is a big part of their body. Are you wondering if your snail is a boy or girl? Well, you're right. It is a boy AND a girl. Snails are both!

Procedure:

[illegible]

1. [illegible]
2. Do [illegible] best [illegible] the [illegible]
3. [illegible]

Possible Ans[illegible]

[illegible] the soft [illegible] foot [illegible]

For More La[illegible]

1. Have a snail [illegible] you [illegible]
2. Hold snail [illegible] your snail [illegible] the [illegible] distance [illegible] such as the minute [illegible]

NAME ______________________________ DATE ________________

Mollusk Lab #2: SNAIL ANATOMY- pg. 1

1. On the list below, draw a line from the body part to the job it does.
2. Label the diagram using the parts listed.

☐	eyes	movement
☐	shell	eating
☐	foot	smelling
☐	tentacles	seeing
☐	mouth	protecting the soft body

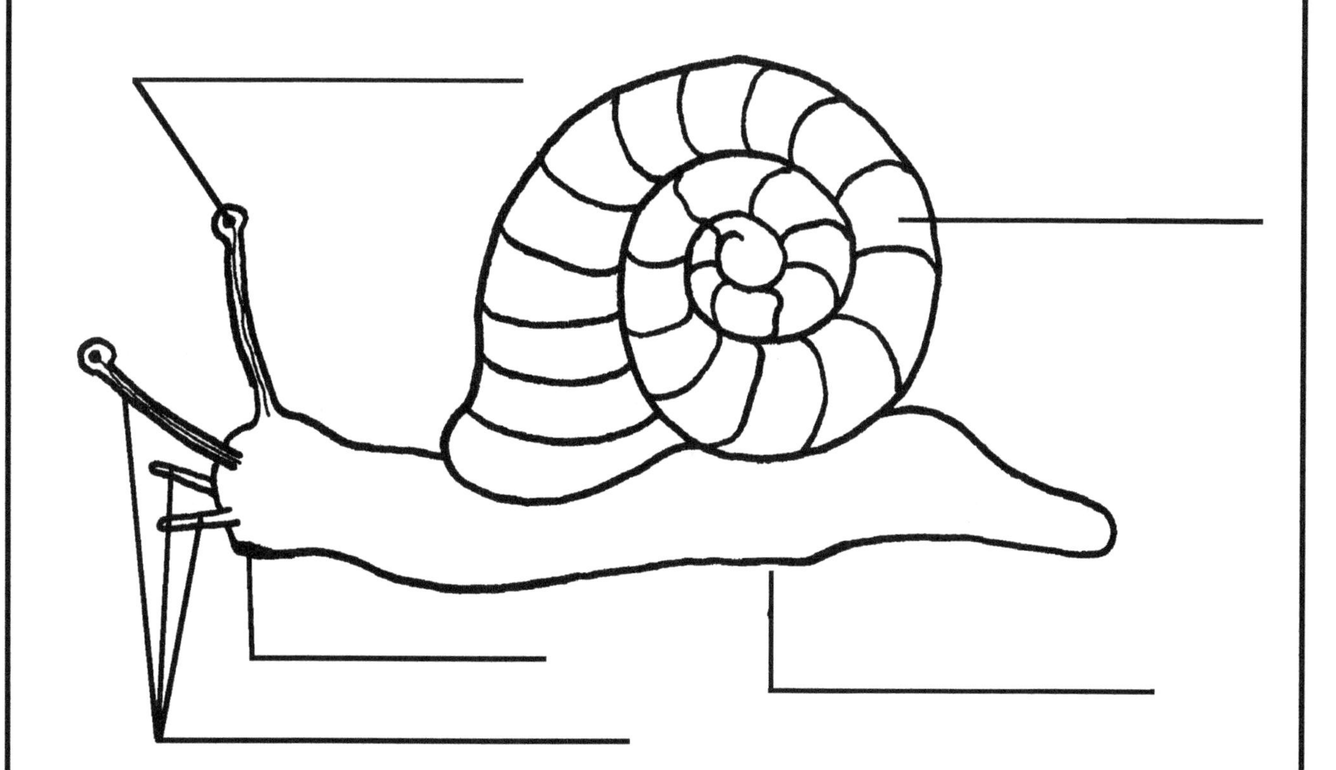

3. In the boxes above, check off each body part as you find it on your own snail.

Mollusk Lab #2: SNAIL ANATOMY- pg. 2

4. Using your centimeter ruler and scale, measure your snail.

Its shell is _________ cm. from front to back

Its body is _________ cm. long

Its antennae are _________ cm. long

It weighs _________ grams.

5. My snail's name is ____________________. It looks like this:

6. Snails have a unique way of eating. They have a special tongue called a radula (rad-you-luh) with hooks on it. Instead of biting, they grate their food, like a cheese grater. Place some food items in front of your snail and listen for grating noises. See if you can discover what your snail likes to eat.

MY SNAIL LIKES	MY SNAIL DOESN'T SEEM TO LIKE

ANIMAL KINGDOM BOOK: MOLLUSKS - instructions

Material:

copy of lab sheet (1 page), pencil or pen
Animal Kingdom Book (assembled)
scissors
glue gel (Elmer's School Glue Gel)
art supplies- markers, colored pencils, stickers, rubber stamps etc.
The Usborne Illustrated Encyclopedia of The Natural World or other animal encyclopedia (optional)

Aloud: Today you are going to add a summary of mollusks to your Animal Kingdom book. Once this book is done you will have a great way to look up information on different animals that you have learned about.

Procedure:

Lab Day:

1. Use the information from your science notebook to fill in sections of your lab paper. List the characteristics of mollusks, examples of mollusks and in the trivia section write the one thing about them that you thought was the most unusual or interesting. You may want to use the internet or *The Usborne Illustrated Encyclopedia of The Natural World* to find a really neat fact about them.
2. Color the animals and the large mollusk label on the page.
3. Page 3 in your Animal Kingdom Book is for your mollusk summary. Cut out the small mollusk label (number 1). Have your book open to page 1 (cnidaria). The mollusk label will go on the small tab for page 3 that shows at the top and to the right of the worm label. Glue your mollusk label in place.
4. Turn to page 3. Cut out all remaining sections of your lab paper. Arrange them on page 3 of your book in whatever order you enjoy. Glue them in place.
5. Finish decorating this page as you wish.
6. Enjoy referring to your book to review what you have learned about the animal kingdom.

Possible Answers:

1. Characteristics of mollusks:
   - soft body
   - hard shell (usually)
   - complex organs
   - mollusks have eyes
2. Examples of mollusks:
   - snail
   - octopus
   - clam
   - oyster
   - squid
   - slug

NAME ______________________ DATE ______________

ANIMAL KINGDOM BOOK- MOLLUSKS

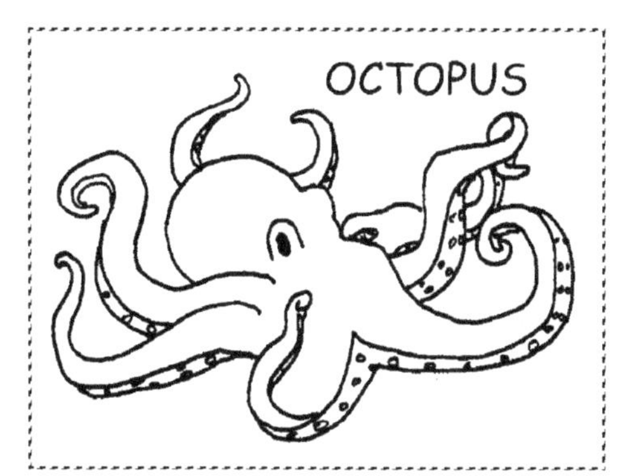

1. MOLLUSKS

2. MOLLUSKS

CHARACTERISTICS OF MOLLUSKS

____

____

____

SNAIL

STRANGE
MOLLUSK TRIVIA

____

____

____

____

____

____

EXAMPLES OF MOLLUSKS

____

____

____

____

____

CLAM

NAME ______________________________ DATE ____________________

For my notebook

ECHINODERMS HAVE SPINY SKIN

Picture an animal creeping along the ocean on hundreds of tube feet, its body covered in spines and its 5 arms so strong it can open a clam with no tools. Now for the really creepy part. When it eats, its stomach moves slowly out of its own body and right into the clam shell to digest its food there. Sounds like a monster from a movie, doesn't it? What I just described is a real sea star, commonly called a starfish. A sea star doesn't seem that fearsome. You may have even held one before.

Sea stars are in a group of animals called <u>echinoderms</u> (eh-**kie**-no-drmz). In latin, "echino" means spiny and "derm" means skin. Can you guess what all echinoderms have in common? They all have spiny skin. A <u>sea urchin</u> (r-chin) is an echinoderm that looks like a pincushion full of long needles. A <u>sand dollar</u> is also an echinoderm. Most sand dollars you see are dead and the short, fuzzy spines have rubbed off. Not only do all echinoderms have spiny skin, most have bodies that can be divided into 5 parts. Most sea stars have five arms but some have 20. Can you divide 20 arms into five even groups? Try this with cubes to find out. Also, if you look at the top of a sand dollar you will see its 5 star shape.

So, next time you are in the ocean and you see a spiny skinned, five armed creature creeping up behind you on hundreds of feet, don't panic - it's no sea monster. It's an echinoderm!

Echinoderm Lab: GIVE ME FIVE - instructions

Material:

lab sheet (1 page), pencil
Weird Echinoderm Facts page (pg. 166)
unpeeled banana knife

Aloud: You heard that every echinoderm has spiny skin. You also heard that their bodies can all be divided into 5 parts. It's easy to see the 5 parts on some. Others are not so obvious. A sea urchin is built like a pincushion and a sea cucumber is built, well, like a big, squishy cucumber. Every one, though, is formed of five rows of muscles. For your lab today you are going to use a banana peel to show how those 5 rows of muscles can form into animals with so many different shapes.

Procedure:

Lab Day:

1. Read over Weird Echinoderm Facts together.
2. Cut the top stem off of the banana.
3. Carefully split the peel into 5 fairly even sections, leaving them attached at the bottom. Eat the banana.
4. Use the banana peel to make the shapes indicated in the lab.
5. Complete the lab paper.

Possible Answers:

1. Sea star or brittle star.
2. Sea cucumber
3. Sea urchin
4. A sand dollar has a 5 star pattern on top. It also has 5 rows of muscles. (It is really just a flattened sea urchin).
5. A sea urchin is formed from 5 arms (muscles) curled over its back and sealed together.
6. A brittle star has 5 arms.
7. A sea cucumber is made of 5 rows of muscles curled over its back in a tube shape and sealed together.

Conclusion / Discussion:

1. How do the spines help these animals survive?

For More Lab Fun:

1. Visit a marine aquarium where you can see, touch and feel echinoderms for yourself.
2. Go to tidepools and find these animals in the wild. Before you do though, learn the local laws concerning marine animals. Also, find out what in the area can be dangerous to touch. Make sure to always handle animals gently and to leave everything as you found it.

NAME ______________________________ DATE __________________

Echinoderm Lab: GIVE ME FIVE

With your banana peel, make the shapes you see below. For the second and the third shapes, picture the centers "filled in". Under each banana peel shape write the name of the echinoderm it looks most like.

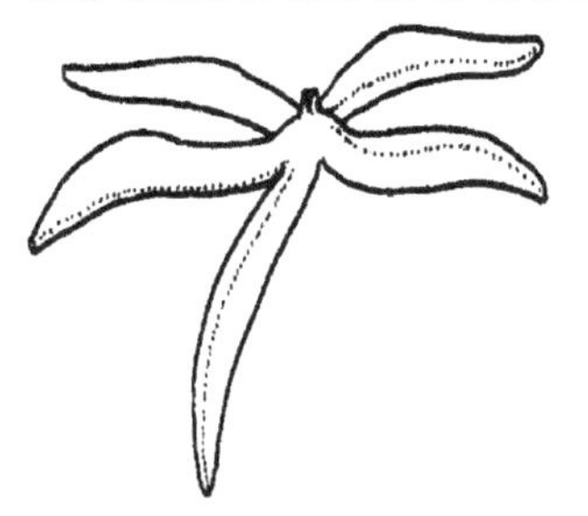

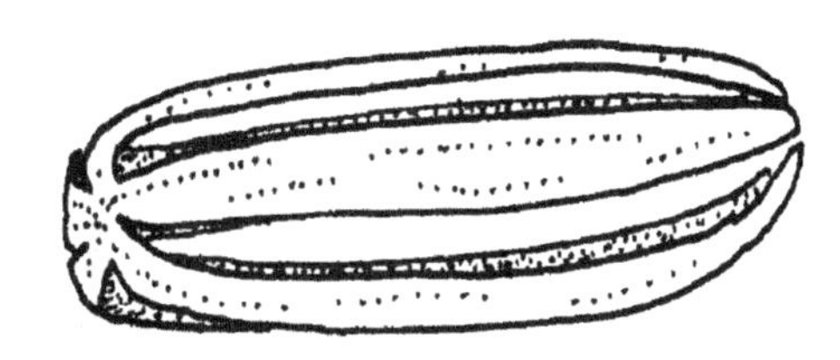

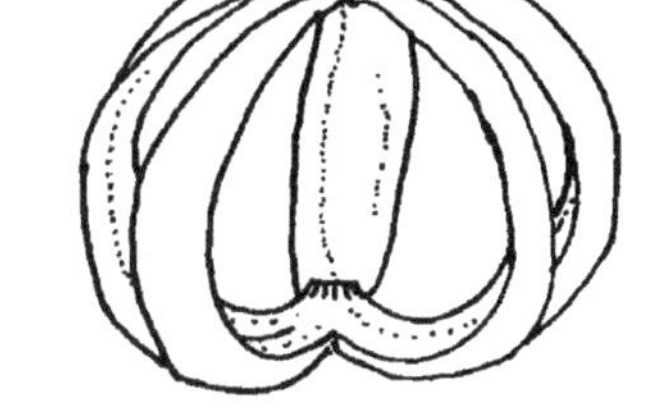

1.__________________ 2.____________________ 3.____________________

What part of each echinoderm is in fives? Some you can tell by looking. For the others, find the information on the Weird Echinoderm Facts page.

Sand dollar

4. ____

____

Sea urchin

5. ____

____

Brittle star

6. ____

____

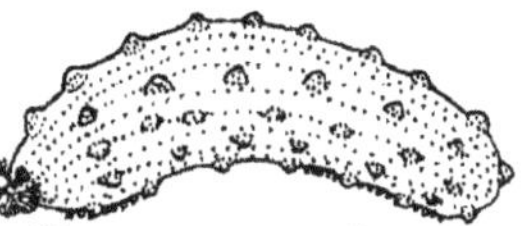

Sea cucumber

7. ____

____

Sea star

8. ____

____

Weird Echinoderm Facts

The largest of sea cucumbers can grow to more than 3 feet long and 6" wide. Measure that!

Sea cucumbers have five long strips of muscles that run the length of their bodies (just like an inverted banana peel) (hint!)

A scared sea cucumber will frighten off its attackers by spitting out its own stomach and intestines. They can then regrow them over time.

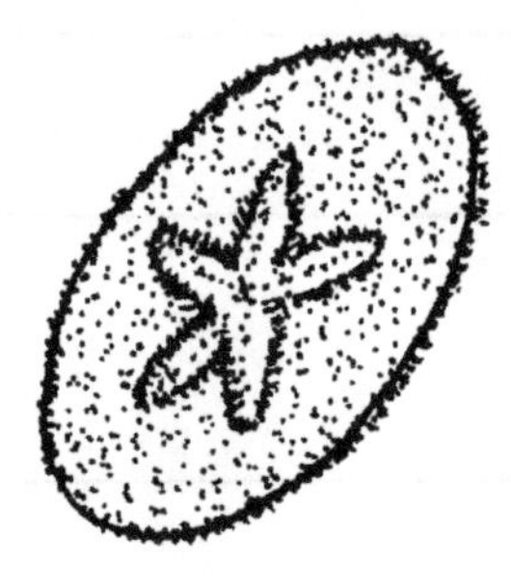

Sea urchins and sand dollars are also made of five arms but the arms have curled up and sealed together over the top of the body to form a ball shape.

Sea urchins have mouths that are in five parts

Sea urchin

The largest sea star grows to over 4 1/2 feet across. The smallest less than 1/4 inch.

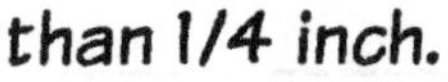

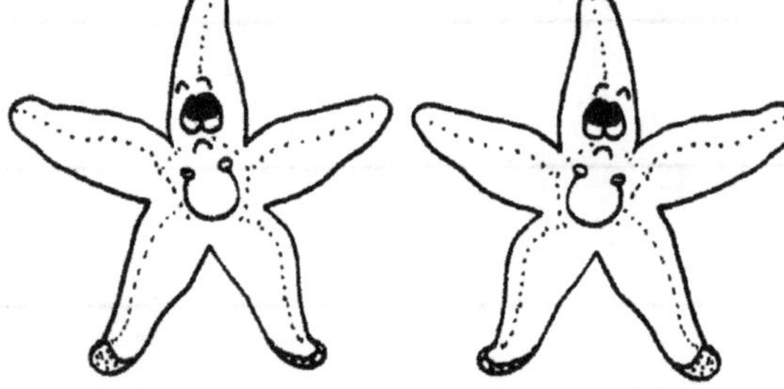

If a sea star loses an arm, it can regrow it. In fact, sometimes a sea star cut into two parts can become two new sea stars!

!

ANIMAL KINGDOM BOOK: ECHINODERM - instructions

Material:

copy of lab sheet (1 page), pencil or pen

Animal Kingdom Book (assembled)

scissors

Glue gel (Elmer's School Glue Gel)

art supplies- markers, colored pencils, stickers, rubber stamps etc.

The Usborne Illustrated Encyclopedia of The Natural World or other animal encyclopedia (optional)

Procedure:

Lab Day:

1. Use the information from your science notebook to fill in sections of your lab paper. List the characteristics of echinoderms, examples of echinoderms and in the trivia section write the one thing about them that you thought was the most unusual or interesting. You may want to use the internet or *The Usborne Illustrated Encyclopedia of The Natural World* to find a really neat fact about them.
2. Color the animals and the large echinoderm label on the page.
3. Page 4 in your Animal Kingdom Book is for the echinoderm summary. Cut out the small echinoderm label (number 1). Have your book open to page 1 (cnidaria) The echinoderm label will go on the small tab for page 4 that shows at the top and to the right of the mollusk label. Glue your echinoderm label in place.
4. Turn to page 4. Cut out all remaining sections of your lab paper. Arrange them on page 4 of your book in whatever order you enjoy. Glue them in place.
5. Finish decorating this page as you wish.
6. Enjoy referring to your book to review what you have learned about the animal kingdom.

Possible Answers:

1. Characteristics of echinoderms:
   - spiny skin
   - tube feet
   - body in patterns of 5's
2. Examples of echinoderms:
   - sea star
   - sea cucumber
   - brittle star
   - sand dollar
   - sea urchin

NAME ________________________ DATE ________________

ANIMAL KINGDOM BOOK - ECHINODERM

1. ECHINODERM

2. ECHINODERM

WOW!
ECHINODERM TRIVIA

____

____

____

____

____

____

SEA STAR

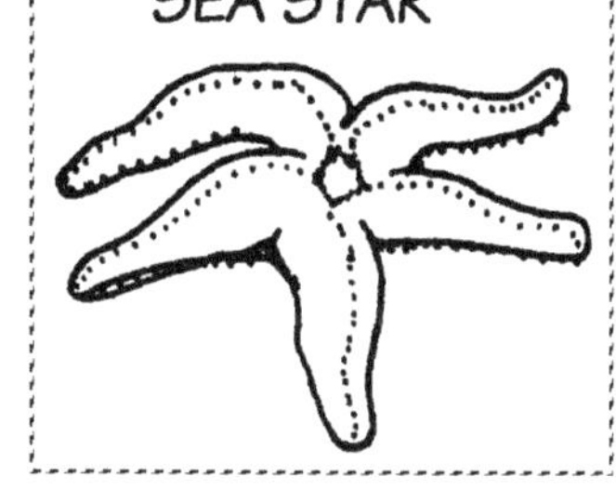

CHARACTERISTICS OF ECHINODERMS

____

____

____

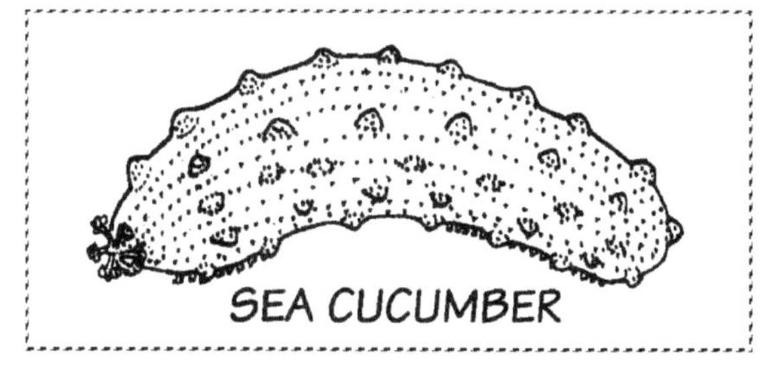

SEA CUCUMBER

SAND DOLLAR

EXAMPLES OF ECHINODERMS

____

____

____

____

____

NAME ______________________ DATE ______________

For my notebook

ARTHROPODS HAVE STIFF LEGS WITH JOINTS

Almost everywhere you go you are surrounded by arthropods (**ahr**-throe-podz). Where are they? WHAT are they? I'll give you a hint. "Arthro" means joint and "pod" means leg, so arthropod means "jointed legs." Look at the legs on a butterfly, a lobster or a spider and you'll see that you have discovered an arthropod. Insects are the biggest group of arthropods. The other major groups of arthropods are arachnids (uh-**rack**-nidz) which includes spiders and crustaceans (kruh-**stay**-shuhnz) which includes lobsters and crabs. You can tell which group of arthropod you have by counting its pods...er legs. Insects have 6, arachnids have 8 and crustaceans have 10 or more. Besides having jointed legs, arthropods also have a hard outside skeleton to protect them. A lobster's shell is its exoskeleton.

So, how about "bugs"? Many people call any kind of creepy critter a bug but most of them aren't really bugs. Scorpions, spiders and even lady "bugs" aren't really bugs. The truth is, bugs are just one group of insects. Do you know what a water strider is? It's that skinny, long-legged insect that scoots around on the top of the water. It IS a bug. So even though they may bother you, mosquitoes and flies really won't "bug" you.

For my notebook

INSECTS HAVE SIX LEGS

When you were a little baby, you looked different than you do now. Besides being smaller, your legs and arms were shorter and your head was a much bigger part of your body. Even though you looked different, except for the size of your parts, you were basically a small version of the person you are now. Not an insect. Their babies look so different it's hard to tell that the baby and the parent are even the same type of animal. Does a caterpillar look like a butterfly? No, not at all. When an animal changes from one form to another, it's called metamorphosis. Some insects, like butterflies, go through four forms. They start out as an egg, hatch into a caterpillar, form a chrysalis and then emerge as a butterfly. Others go through three changes from an egg, to a small, adult-like version to an adult. So, with so many different kinds of insects, and all with different body forms, how do we know that what we have found is an insect? Besides having an exoskeleton, every insect has six legs and three body parts. They also usually have two pairs of wings.

Next time you find an insect, don't grab for the swatter. Reach instead for a magnifying glass and get a closer look at this amazing, form-changing arthropod.

Insect Lab #1: BUTTERFLY METAMORPHOSIS - instructions

****NOTE #1: Plan this activity for the spring or summer. Most companies will not ship butterfly larvae in the winter as the butterflies can not be released then.**

****NOTE #2: This lab can be done simultaneously with Insect Lab #2. This will save you from having to find or purchase two sets of butterfly larvae (caterpillars).**

Material:

copy of lab sheets (2 pages), pencil
* butterfly larvae (caterpillars)- must be ordered ahead of time. See materials list for ordering info. You may choose to find and catch your own. Larvae can be kept in a box with plenty of leaves from the EXACT plant they came off of. They are voracious eaters and may need to be fed every day.
*butterfly house- order ahead or make your own from a big box with window cut in. Tape cellophane in window so you can see your butterflies.
centimeter ruler
hand lens (optional)

Aloud: Butterflies go though complete metamorphosis. This means they have four very different body shapes in their lives. During metamorphosis, a butterfly starts as an egg, hatches into a caterpillar, eats constantly and becomes a chrysalis. Inside the chrysalis, the caterpillar goes through some amazing changes to emerge as a beautiful butterfly- it's adult form. The adult butterfly will then lay more eggs and the whole cycle starts again. Other animals go through metamorphosis too. Scientists call the stages of complete metamorphosis the egg, larva, pupa and adult. In a butterfly, these stages are the egg, caterpillar, chrysalis and adult butterfly.

Procedure:

AT LEAST THREE WEEKS BEFORE THE LAB:

1. Make sure to order your butterfly house. They should come with a certificate for butterfly larvae, which must be ordered separately.

AT LEAST ONE WEEK BEFORE THE LAB

1. Order your butterfly larvae. They will come in a container with all the food you need.

Day 1: WHEN LARVAE ARRIVE:

1. Draw larva in as much detail as possible. Count legs (6 true legs and 10 false legs) and wings (0). Measure length and width, complete section one of lab, except last question.

Day 5 or so: WHEN LARVAE ARE BIG AND EASY TO SEE

1. Go to Insect Lab #2- pg. 1

Day 8 or so: WHEN FIRST PUPA FORMS

1. Record the day #. Draw the pupa as accurately as possible. Count legs (hard to see) and wings (2) and measure pupa. Complete section 2 of lab, except last question.
2. Figure out how long the larva stage lasted by adding a couple of days (time in the mail) to how long you had him. Record in section 1, last question.

Day 18 or so: WHEN FIRST BUTTERFLY EMERGES

1. Record the day #. Draw and measure butterfly. Count legs (6) and wings (4). If you are using Painted Lady butterflies it may appear as if they only have 4 legs. Their front legs are modified into tiny "pro-legs" which are held against the underside of the body near the "mouth". Complete section 3.
2. Fill in the last question of section 2.
3. Go to Insect Lab #2- pg. 2.
4. Butterflies can be released now or kept for egg laying. If you choose to have a house full of larvae (and you really might!) see "For More Lab Fun" on page 174. If they lay eggs, do section 4 of the lab. If not, you can find the information on eggs on the internet or in the Conclusion below.
5. Color the graph. Answer the questions.

CONTINUED ON BACK

Insect Lab #1: BUTTERFLY METAMORPHOSIS - instructions - pg. 2

Possible Answers:

3. Butterflies do all their growing when they are caterpillars (the larva stage).
4. Larvae spend almost all their time eating. Some children may not see this so you might want to point out how much food disappeared during this life stage.
5. Most changes in shape and appearance occur in the chrysalis (pupa stage).
6. Well, it looks like they're resting but they are actually spending their time and energy changing.

Conclusion / Discussion

- Painted Lady butterfly eggs are about 1 mm. long. They are a pale green and oval shaped.
- The entire butterfly life cycle can vary greatly in length depending on the temperature. If you keep them inside in the range of 70° to 90° this is about how long things should take:

Egg	-	3 - 5 days
Larva	-	5 - 10 days (4 weeks or more outside) Pupa
	-	7 - 10 days
Adult	-	about 5 days to lay eggs and can live about 2 weeks

For More Lab Fun:

1. If you keep the adult butterflies, they will breed about 2-3 days after emerging from the chrysalis. In another few days to 2 weeks they will lay eggs and you can start the entire life cycle over again. They must have a good supply of hollyhock, Malva, nettle or thistle leaves to lay their eggs on and also for the larvae to eat. They can tell the difference and won't lay their eggs on just any old leaf.
2. Keep a daily journal of growth and changes. Draw and measure your caterpillar daily.
3. Mark the level of the food at the bottom of the container or weigh the leaves as you put them into the container. Figure out how much food a caterpillar eats compared to its own size.
4. If a human baby grew as fast as a caterpillar, he would weigh six tons in two weeks. Find something that weighs six tons (12 horses) and compare that to a two week old human infant.

NAME ______________________________ DATE ____________________

Insect Lab #1: BUTTERFLY METAMORPHOSIS- pg. 1

DAY 1: My butterfly LARVA	...is ______cm. long and ______cm. wide. I can count _____ legs and _____ wings I would describe my larva as: ______________________________ It stayed a LARVA for _____ days.
DAY ____: My butterfly PUPA	...is ______ cm. long and______cm. wide. I can see _____ legs and_____ wings I would describe my pupa as: ______________________________ It stayed a PUPA for ______ days.
DAY ____: My ADULT butterfly	_____cm. long and ______ cm. wide. I can count _____ legs and _____ wings I would describe my butterfly as: ______________________________ My butterfly laid eggs on day______ or... I released my butterfly on day _______
Butterfly EGGS	OPTIONAL: My butterfly eggs: length ______ mm. color ___________ shape_____________ They took _____ days to hatch. **I'm a grandparent!**

Insect Lab #1: BUTTERFLY METAMORPHOSIS- pg. 2

Color on the graph below how long each stage of the butterfly life cycle lasted.

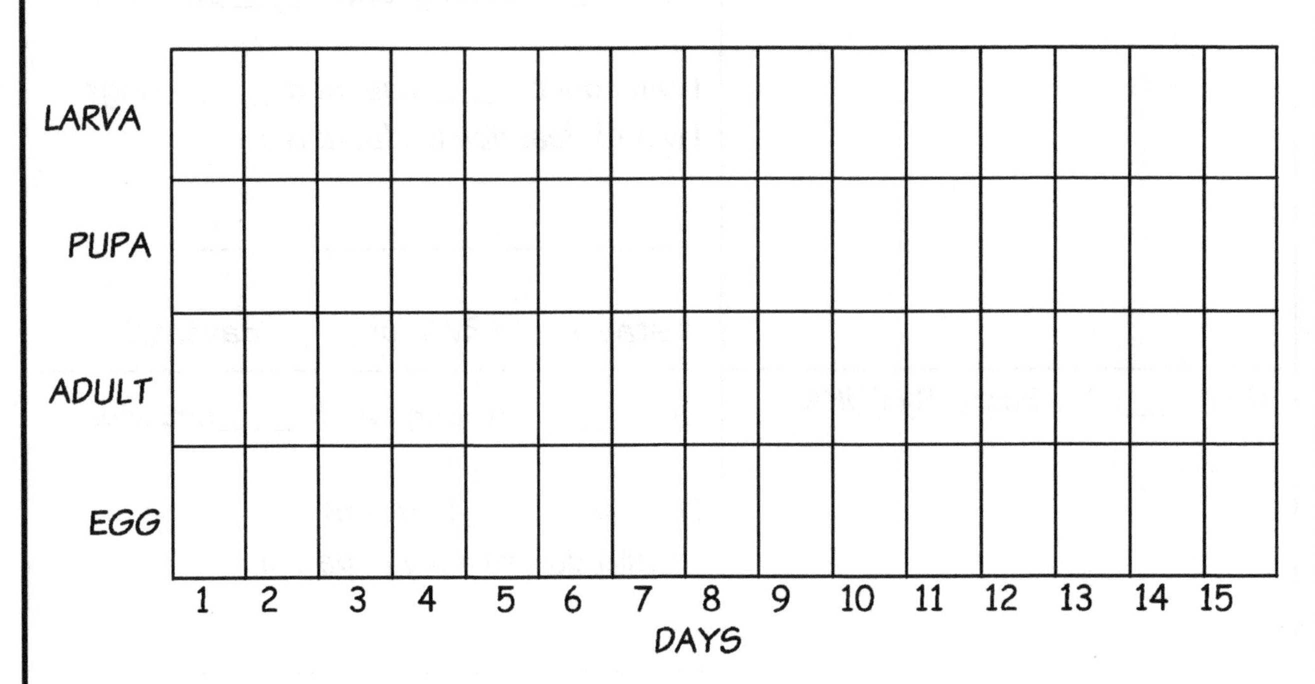

1. Which stage of metamorphosis lasted the longest? ________________

2. In days, the entire life cycle of my Painted Lady took:

   Larva ____ + Pupa ____ + Adult ____ + Egg ____ =

   ________ total days for the entire life cycle

3. My Painted Lady changed the most in size during the ___________ stage.

4. It spent most of its time __________________ during this stage.
   (Doing what?)

5. My Painted Lady changed the most in appearance when it went from the

   ______________stage to the ______________stage.

6. It spent most of its time ________________ during this stage.
   (Doing what?)

Insect Lab #2: CATERPILLAR TO BUTTERFLY - instructions

****NOTE: This lab can be done simultaneously with Insect Lab #1. This will save you from having to find or purchase two sets of butterfly larvae (caterpillars).**

Material:

copy of lab sheets (2 pages), pencil
butterfly larvae (caterpillars) and butterfly house- see Insect Lab #1 for ordering instructions
colored pencils or skinny black marker and various colored crayons

Aloud: There is a tremendous variety in the insect world, but even though an ant doesn't look much like a butterfly or a cockroach, they all have some things in common. As you watch your butterfly larva develop into a pupa and then into an adult butterfly, you will see many differences, even though it is the very same insect! By comparing your caterpillar to your butterfly you will be able to see many differences between the two, and also many features that they share with every other insect.

Procedure:

Lab Day 1: WHEN CATERPILLARS ARE BIG ENOUGH TO SEE CLEARLY

1. Compare your caterpillar to the one in the diagram (pg. 1). Label and color the diagram as indicated. On your own larva, locate as many of the parts listed as you can.

Lab Day 2: WHEN BUTTERFLIES EMERGE

1. Compare your adult butterfly to the one in the diagram. Label and color the diagram as indicated. On your live butterfly, locate as many of the parts listed as you can.

Possible Answers:

1. Page 1, Caterpillar diagram:

1. Head	2. Thorax	3. Abdomen	4. Antennae
5. True legs	6. Setae	7. False legs	8. Spiracles

2. Page 2, Butterfly diagram

1. Head	2. Thorax	3. Abdomen	4. Antennae
5. Legs	6. Fore wing	7. Spiracles	8. Hind wing

For More Lab Fun:

1. During the adult (butterfly) stage, line the bottom of your container with black paper. Place circles of different colors on the floor. Record which colors the butterflies go to more. Answer: They prefer yellow and white then blue, orange and pink. They don't seem to see red at all.
2. Order two batches of caterpillars. Keep one set at a higher temperature than the other (but no higher than 100° F and no lower than 60° F). Compare rates of change from larvae to pupae and pupae to adult. Answer: The colder they are, the longer they take to go from one stage to the next.

For Your Information:

Larva is singular	–	Larvae is plural
Pupa is singular	–	Pupae is plural
Antenna is singular	–	Antennae is plural

NAME ______________________________ DATE ______________________________

Insect Lab #2: CATERPILLAR TO BUTTERFLY- pg. 1

Use the clues to label the diagrams below.

Insects have 3 body parts.

The first section is the head, Write head in the box above the head.

The part where legs attach is the thorax.

The sections with little breathing circles make up the abdomen.

Insects have 6 true legs. Your caterpillar also has 10 false legs. Label both kinds of legs.

Insects breathe through spiracles (**speer**-uh-klz), tiny holes along their bodies. Label the spiracles.

Some insects have bristles to protect themselves from predators. Label these setae (**see**-tee)

Insects have antennae. On your caterpillar they are small and lie against the head. Label the antennae.

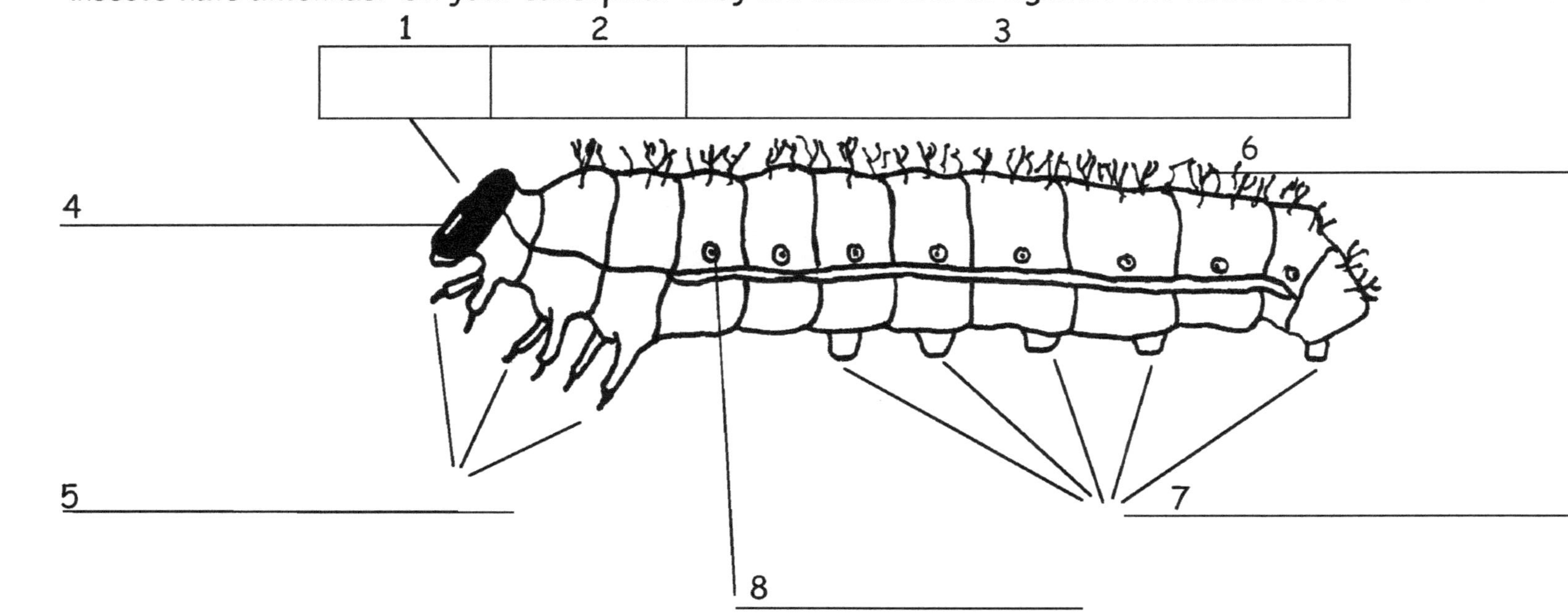

Color the caterpillar to look like the one you are raising.

Insect Lab #2: CATERPILLAR TO BUTTERFLY- pg. 2

Use the clues to label the diagrams below.

Insects have 3 body parts.

The first section is the head, Write "head" in the box to the left of the head.

The part where legs attach is the thorax.

The sections with little breathing circles make up the abdomen.

Insects have 6 legs. The two front legs of the Painted Lady are small and folded up near the head.

Insects breathe through spiracles (speer-uh-kulz), tiny holes along their bodies. Label the spiracles.

Many insects have 4 wings. Hind means back and fore means front. Label the hind wings and fore wings.

Insects have 2 antennae. Label the antennae.

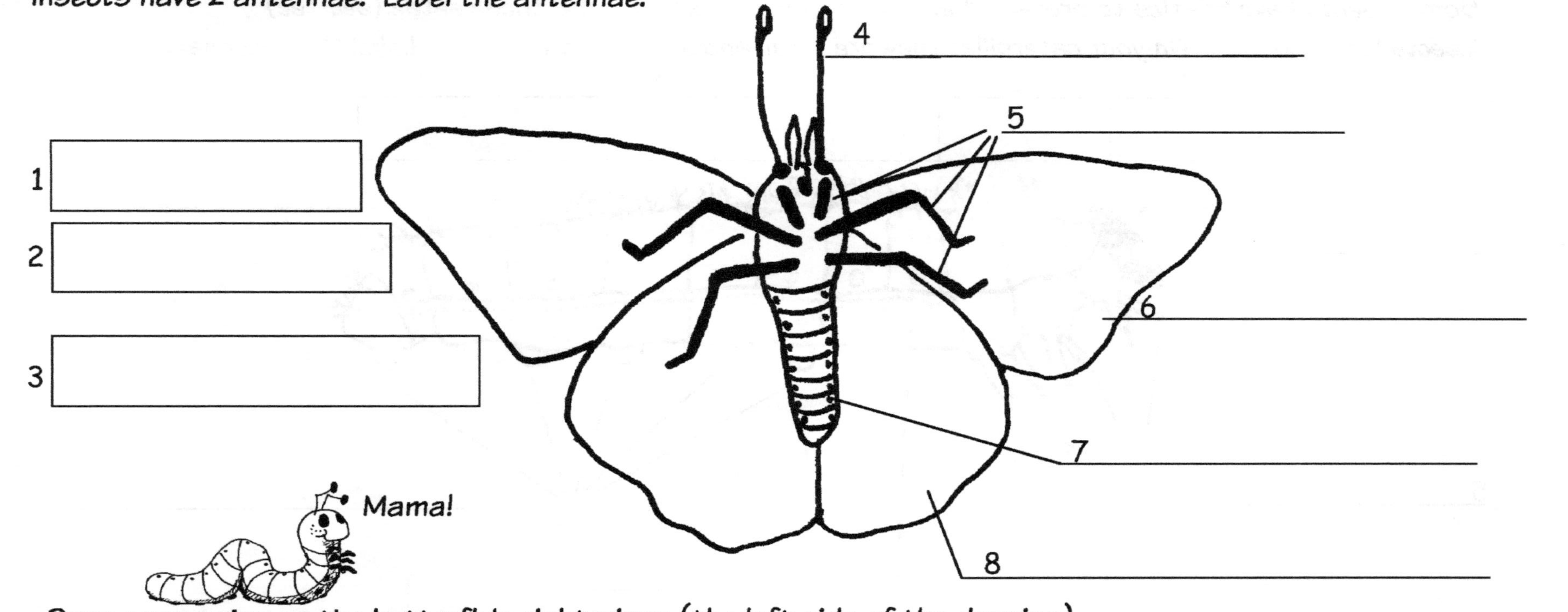

Draw some veins on the butterfly's right wings (the left side of the drawing).

Color that side of the butterfly to match the one you are raising.

NAME ______________________ DATE ______________

For my notebook

SPIDERS ARE ARACHNIDS

Have you ever looked really closely at a spider? Go find a picture of one, or better yet, a live one to look at. If you find a live one, put it in a clear container so you can see it well. Before you catch a live one, ask a parent first. There are two dangerous spiders common to the United States (the Black Widow and the Brown Recluse) and you don't want to touch one of those. Now, take a good, close look at your spider or picture. If you remember, insects have 6 jointed legs, 3 body parts and usually 2 pairs of wings. Spiders have jointed legs just like insects, don't they? This means they are arthropods also. Did you remember that arthropod means "jointed legs"? Also, like insects, spiders have a hard exoskeleton and no bones inside. Unlike insects though, spiders have 8 legs, 2 body parts and no wings. This means they are not insects and definitely not bugs. They are in a group called <u>arachnids</u> (uh-**rack**-nidz). Other 8 legged arthropods are scorpions and ticks. You probably never thought that ticks were related to spiders, but they are. They are arachnids too.

Let's learn a song to help us remember why spiders and insects are in different classes. In the meantime, enjoy looking at your new buddy and have fun getting to know arachnids a little better!

NAME ____________________ DATE ________________

THE SPIDER SONG

sung to the tune of "Oh My Darling Clementine"

It's a spider, not an insect.
It has eight legs instead of six.
A head and body,
but no wings,
And its relatives are ticks.

Draw a picture of your spider below. Be as accurate as possible.

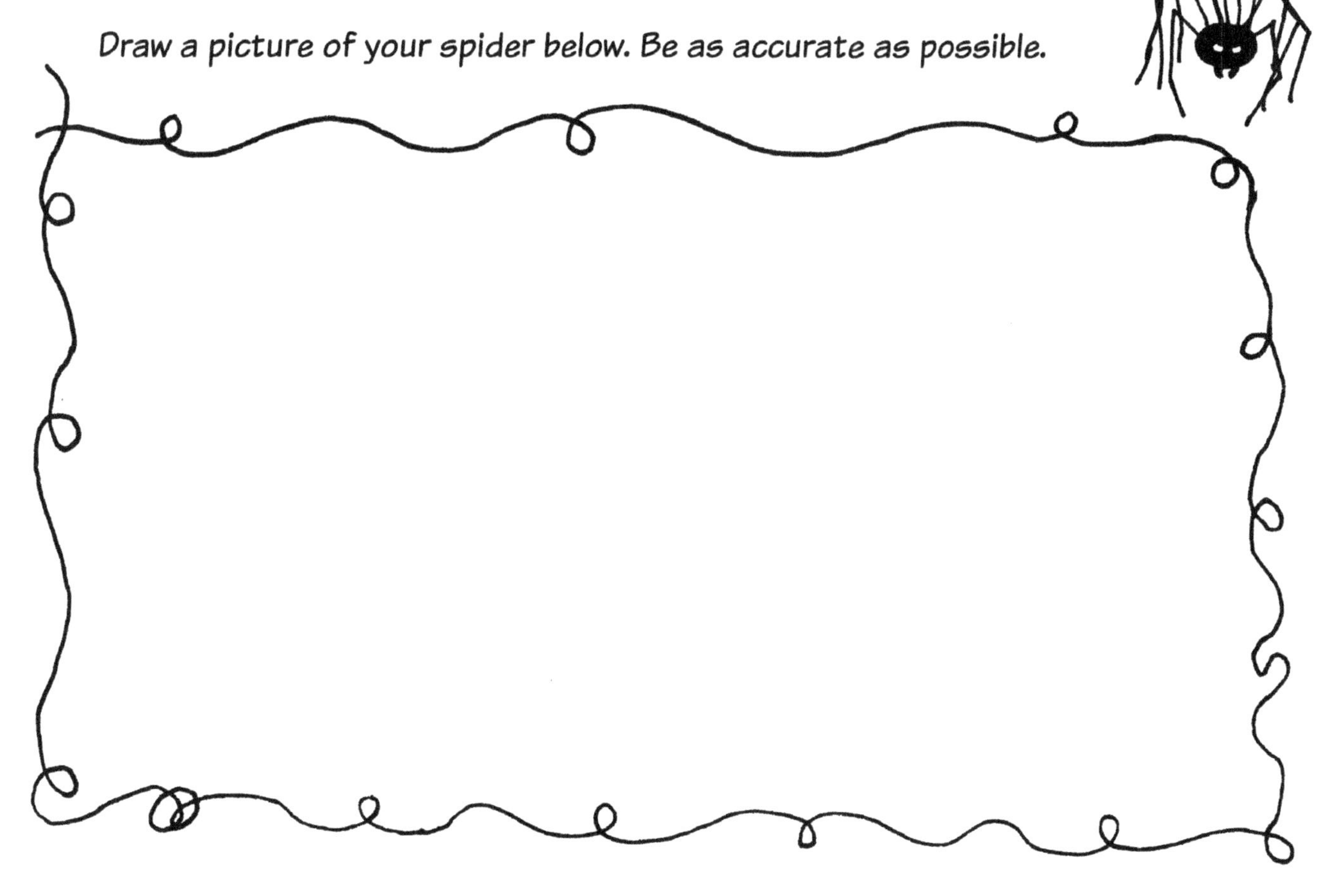

Arachnid Lab: INSECT OR SPIDER? - instructions

Material:

copy of lab sheets (2 pages), pencil
pair of scissors
glue gel (Elmer's School Glue Gel)
pictures of spider and insect (or the real thing!)

Aloud: All arthropods have jointed legs and an exoskeleton. Some arthropods have six legs and some have eight. Do you remember which ones have 8 legs? Arachnids, right? Spiders are arachnids. They have 8 legs and 2 body parts. How many body parts does an insect have? Three. Many insects also have 4 wings. Let's put together an insect and a spider to show the differences between them.

Procedure:

Lab Day:

1. *Review the differences between insects and spiders. Insects have 6 legs, 3 body parts and usually 4 wings. Spiders have 8 legs, 2 body parts and no wings. Both are arthropods with jointed legs and an exoskeleton.*
2. *On lab pg. #1 decide which outline is the insect and which is the spider.*
3. *Cut out the legs and wings from lab pg. #2. Cut along the dotted lines so you don't have to go around each leg.*
4. *Glue legs and wings onto the proper body outline. See diagram below for proper placement.*
5. *Fill in answers on lab sheet including additional facts. Examples may be: I (spiders) spin webs. I am an arachnid. I (insects) can fly. I have antennae.*

For More Lab Fun:

1. *Spiders are fun and fairly easy to house for a time. Keep one in a plastic jar with a stick. Feed them flies or other tiny insects and you'll be amazed at how much fun it is to watch them hunt. Jumping spiders are easy to find and fun to keep. They don't need to be watered but should be fed at least a couple of times a week.*

For Your Information:

*See diagrams below of spider (on the left) and insect (on the right) for proper placement of wings and legs. On the insect the three body parts are the head, thorax and abdomen. On the spider the head and thorax are combined together to form a cephalothorax (seh-fuh-low-**thor**-ax) and then the abdomen is the second body part. Legs and wings attach at the spider's cephalothorax and the insect's thorax.*

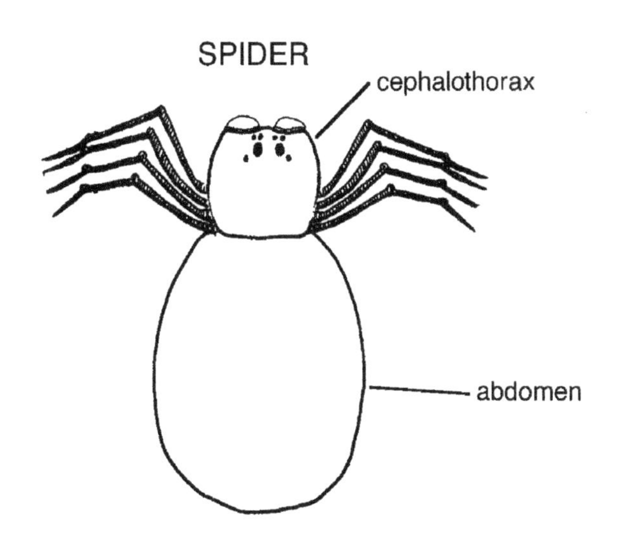

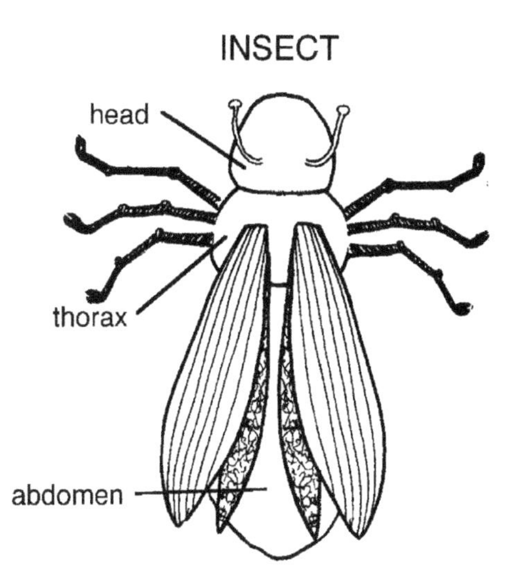

NAME ______________________ DATE ______________

ARACHNID LAB: INSECT OR SPIDER?- pg. 1

I have _____ body parts. That means I am a/an ________________

I have _________ legs.

I have _________ wings.

Other things about me:

____

____

____

____

____

I have _____ body parts. That means I am a/an ________________

I have _________ legs.

I have _________ wings.

Other things about me:

____

____

____

____

____

NAME ______________________________ DATE ____________________

ARACHNID LAB: INSECT OR SPIDER?- pg. 2

Cut out the parts below. To make things easier, don't go around each leg, simply go along the dotted line where there is one. Glue each part on the proper animal on the previous page. When you are done gluing, color in your insect and spider if you wish.

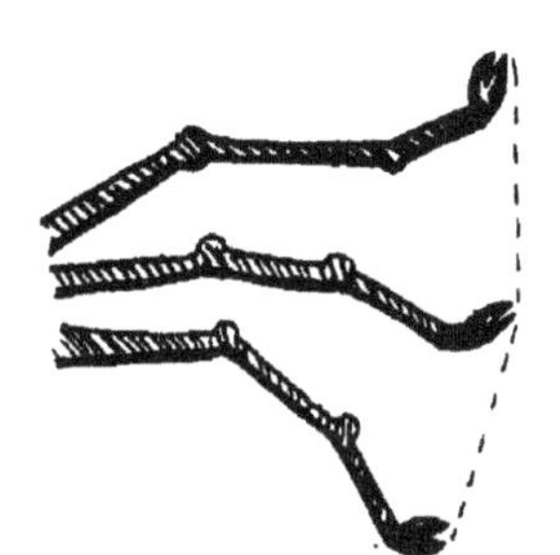

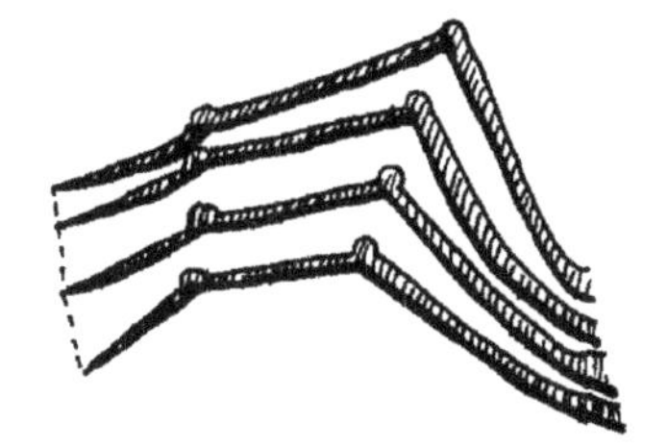

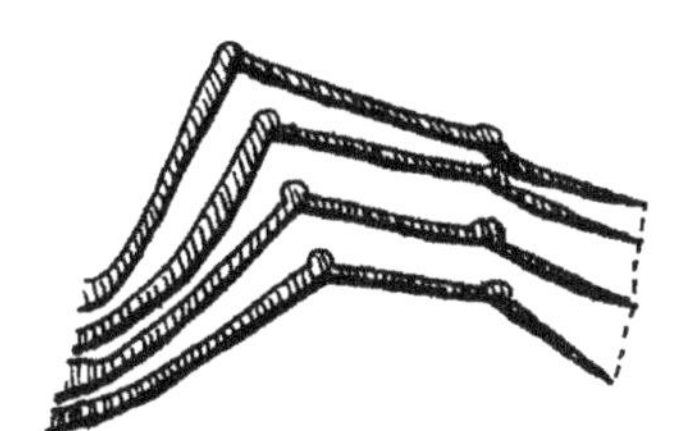

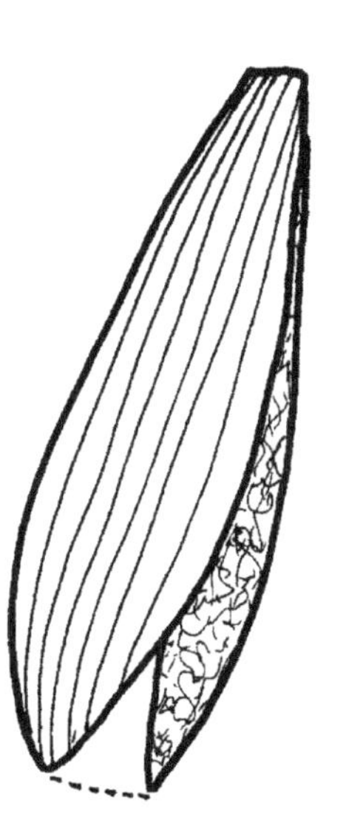

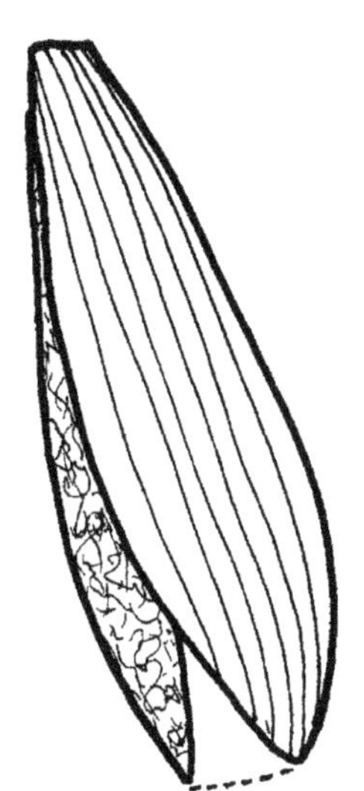

NAME ______________________________ DATE ___________________

For my notebook

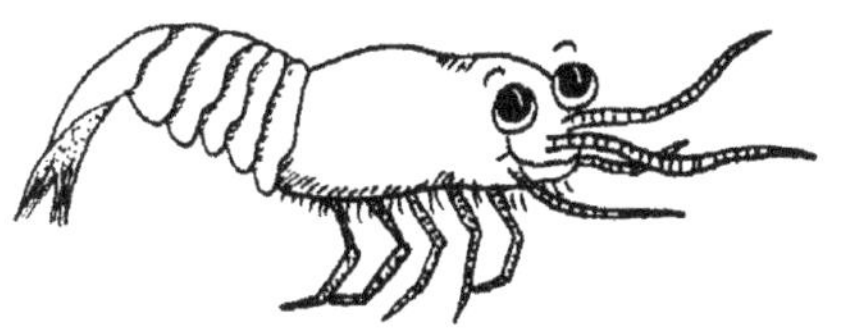

CRUSTACEANS FILL THE SEAS

There are insects all over the world—in fact filling the world from forests to deserts and from lakes to pitch-black caves. Where can you go to get away from insects? The sea. Insects do not live in salty water. There are, though, arthropods that fill the seas like insects fill the land – crustaceans (kruhs-**tay**-shunz). Some of the best known salt water crustaceans are crabs, lobsters and shrimp. A few crustaceans, like crayfish, live in fresh water and an even smaller number live on land. Have you ever played with a pill bug or sow bug? If so, you have met one of the few land crustaceans.

Crustaceans are like all other arthropods in that they have stiff, jointed legs, segmented bodies and an exoskeleton—in some cases a really hard exoskeleton. Where insects have 6 limbs and arachnids have 8, crustaceans have 10 or more. Insects have one pair of antennae, arachnids have none and crustaceans have 2 pairs or 4 antennae.

We're going to be learning more about crustaceans by finding and experimenting with sowbugs. They are very common and very easy to keep. In fact, unlike dogs, cats and hamsters, all they ask for is a little moist dirt, a few rotting leaves and a little spritz of water each day.

Crustacean Lab #1: ISOPOD HUNT - instructions

Material:

- copy of lab sheet (1 page), pencil
- small jar or other small, covered container to keep isopods during the hunt
- critter keeper, food storage container, etc. about 8" x 5" x 3" deep (or bigger)
- moist soil or sand
- piece of rotting log to fit into container
- water mister hand lens

Aloud: There is a type of crustacean that lives on land, doesn't pinch and rolls into a ball when scared. Can you guess what it is? It's a pill bug. We call them roly-polies, sow bugs or pill bugs but they aren't bugs. Bugs are a kind of insect. Pill bugs are a kind of sow bug that can roll into a ball. We are going to call them all isopods (eye-so-podz). Today we are going on an isopod hunt. Isopods are fun to hunt for and neat to keep. You can house them just for the experiment we will do next time or you can keep them longer. All they need is a home with moist dirt and a little rotting food to eat. Make sure to mist them well once a day or they will dry out and die. Crustaceans breathe through gills like fish do. Because these isopods live on land they have to find ways to stay wet so they can breathe. Look for them in wet places, like under small rocks, sheets of plastic and logs.

Procedure:

Lab Day:

1. Set up isopods' home before you go hunting. Put about 1" of moist sand or soil in the bottom of the container. Add a piece of rotting bark or log. If you have no lid, one can be made from waxed paper held on with a rubber band. Whatever lid you choose, punch just a few small holes so moisture doesn't leave too quickly.
2. Take your small container and head for the yard. Look in moist, protected and shady places. Under plastic sheeting, rocks, leaves etc. Remember to replace everything you move so the other inhabitants don't die. You will need at least ten for the next lab so get extra. They could be hard to find in their container.
3. Bring back your isopods, put most of them in their new home. Keep a couple of isopods out to do the lab today.
4. Study your isopods closely with your hand lens. Complete parts one (as best the children can) and two of the lab. Body segments refers to the number of shell sections. Body parts refers to the division of head, thorax and abdomen. Both of these are very hard to distinguish and count. Answers are below. You may tell them the answers and have them confirm this or let them do their best.
5. Do part 3 of the lab.

Possible Answers:

#1 Isopods have 3 body parts (head, thorax and abdomen), 14 body segments (one head, 7 thorax and 6 abdomen), 14 legs, 4 antennae, 2 eyes and no wings. Some of these are hard to see. Remember, counting and observing are the goals, not perfection.

#3-4. Isopods match crustaceans in all ways listed (6) and insects in number of body parts only (1). Help children concluded that isopods really are more like lobsters and crabs than like insects.

CONTINUED ON THE BACK

Crustacean Lab #1: ISOPOD HUNT - instructions page 2

For More Lab Fun:

1. If you choose to keep your isopods for any length of time, add an apple core, small piece of potato or almost any other fruit or vegetable to the container once a week or so. Also, remember to water daily. You are going for moist, not dry and not soggy. You will be amazed at how many isopods you have over time. They are very eager to produce a family.
2. Millipedes, which are round like a pencil, are fun to keep in the same container with your isopods. They are from an arthropod group we didn't study. They are totally gentle and they eat and live just like the isopods. They will also surprise you with youngsters if you keep them long enough. CAUTION: Don't make the mistake of putting in centipedes. They are similar but flattened looking. They will eat your isopods and can give you a painful pinch. If in doubt, leave them out!

For Your Information:

The names isopod, roly-poly, pill bug and sow bug all rolling around in your head now? They are often used interchangeably but to summarize: The group arthropod (jointed legs) includes crustacean (10 or more jointed legs) which includes isopods which includes sow bugs (land isopods), which includes pill bugs, also called roly-polies (which are simply types of sow bugs that can roll up). So, is that as clear as mud? If you say isopod you cover all bases, whether you find a pill bug or some other sow bug—they are similar. Also, if you opt to say isopod, you can wow and amaze your friends and encourage your students to think beyond the overused and improper term "bugs."

NAME ______________________ DATE ______________________

Crustacean Lab #1: ISOPOD HUNT

1. Use your hand lens to look very closely at one of the isopods you have found. In the spaces below, fill in how many of each part you can count.

☐	body parts	☐	body segments (shell parts)
☐	legs	☐	antennae
☐	eyes	☐	wings

2. Using the numbers you counted, draw your best isopod in the box below

3. We call isopods "pill bugs" and "sow bugs" and yet scientists have put them in the crustacean group instead of the insect group. Use the checklist below to decide which group isopods should really belong to. For each characteristic, put a check in the box next to each group isopods match.

ISOPODS		CRUSTACEANS		INSECTS
3 body parts		3 body parts		3 body parts
14 legs		10 or more legs		6 legs
4 antennae (2 are very tiny)		4 antennae		2 antennae
no wings		no wings		usually 4 wings
gills for breathing		gills for breathing		breathe through body
exoskeleton forms hard shell		exoskeleton forms hard shell		exoskeleton- no shell

4. Isopods match crustaceans in _______ ways and match insects in ______ ways.

Based on this information, I think isopods should belong in the ______________________ group.

Crustacean Activity: ROLY-POLY POETRY - instructions

Material:

copy of activity sheet (1 page), pencil
colored pencils etc. for decorating (optional)
chalk board, white board or butcher paper

Aloud: *Think of all the names people have for isopods. How many can you think of? For each species, or type of isopod there is only one scientific name but there are many common names. Today you are going to write a poem about your isopods. Try to think of creative but accurate words to describe your isopod in your poem. If you come up with a really great poem, share it with your grandparents and your friends.*

Procedure:

1. Brainstorm words that describe isopods. Write your list on the chalkboard or on butcher paper. Make sure to think of verbs as well as nouns and adjectives. Also, brainstorm other names for isopod. It's amazing we can have so many names for such a small critter.
2. Use the formula given on the activity page to write your cinquain (sing-**kayn**). Use the list of isopod terms you have made to help.
3. Once you have the hang of it, do a second poem, if you wish. Try to use two completely different names for "isopod."
4. Decorate around your poem.

For More Lab Fun:

1. Rewrite your poem onto poster paper and decorate with stickers, drawings, frames etc.

For Your Information:

In case you are poetically challenged (like I am) and have never seen a cinquain, here are 2 examples of cinquains, written by a 4- and a 5-year-old , about snails:

Snail
Gluey, sticky
Climbing, sliding, gliding
My snail feels good
Slug

Snail
Skinny, soft
Sliding, biting, scraping
Snails are soft and mushy
Mollusk

NAME ______________________ DATE ______________________

Crustacean Activity: ROLY-POLY POETRY

Describe your isopod with a cinquain. A cinquain is a poem that follows this pattern:

Line 1: The subject of the poem
Line 2: Two adjectives that describe the subject
Line 3: Three "ing" verbs that describe the subject
Line 4: A 4 word phrase that tells your feelings towards the subject
Line 5: A synonym, or other name for the subject

Line 1: ______________

Line 2: ______________, ______________

Line 3: ______________, ______________, ______________

Line 4: ______________________________

Line 5: ______________

Now, see if you can do another using 2 different names for your subject (isopod).

Line 1: ______________

Line 2: ______________, ______________

Line 3: ______________, ______________, ______________

Line 4: ______________________________

Line 5: ______________

Crustacean Lab #2: A HOME FOR ISOPOD - instructions

Material:

copy of lab sheets, pencil
10 isopods (sow bugs, roly-polies, pill bugs)
tray with sides: 9" x 9" baking pan, food container etc.
heating pad (or hot water bottle)
ice
gallon size zipper-type baggie
water source
paper towels
flashlight

Aloud: Isopods are in a group called crustaceans. Do you remember where most crustaceans live? In salt water, right? They must get their oxygen out of the water, just like you get yours from the air. To do this, they need gills. Fish have gills, don't they? So do crustaceans. This is a problem for isopods since they don't live under water. They must stay wet all the time or they will die. Today we will experiment to find out what environment is best for your isopods. First you will guess how it is best for isopods to stay wet, then you will test your isopods to find out if their behavior helps to keep them alive.

Procedure:

Lab Day: Preparation is required about 15 minutes before lab

1. Set your heating pad on high. Fill the large baggie with ice. Put the heating pad and the ice next to each other without touching. Set the tray on top of both so that half of the tray gets hot and the other half cold. You might have to prop up the heating pad side so the two sides are level. Let set for 15 minutes or so before proceeding with experiment.
2. While tray is heating and cooling, do "My Prediction" part of lab. Guess which direction your isopods will go for each situation. Think about how they can best stay moist to breathe.
3. Place 10 isopods in center of tray, between the hot and cold sides. Watch them for 10 minutes or so and note how many end up on each side. Isopods aren't usually "quick." They often have to "test" each side so waiting is important for best test results.
4. Cover one side of the bottom of the pan with a wet paper towel and the other side with a dry paper towel. Leave a small gap in between so the water doesn't wick across to the dry side. Place isopods on gap and note how many have gone in each direction after 10 minutes or so.
5. Spread a wet paper towel across entire bottom of tray. Cover over 1/2 of the top with foil. Shine flashlight down so it lights up the uncovered side, but doesn't go into the covered end. Place isopods between the dark and light sides and note which side they end up on.
6. Finish lab. Fill in chart on page 2.

Conclusion:

1. Isopods do whatever it takes to stay moist. They will eventually end up on the side of cool, wet and dark. Discuss with the children how this helps them to not dry out.

For More Lab Fun:

1 Experiment with shallow dishes with the same amount of water in each to see which evaporates the fastest. Place one dish in a cool place and another in a warm place. Try a second set with one dish in the sun and another in the shade. Compare your results to where your isopod prefers to "hang out."

2. Do the same experiments but try the isopods one at a time. Do they respond the same? Does the behavior of one seem to influence the behavior of the others?
3. Place 5 isopods on each side of the trays for each test. Do they still end up in the same places?

NAME ______________________________ DATE __________________

Crustacean Lab #2: A HOME FOR ISOPOD - pg. 1

Where do you think your isopods would prefer to call home? For each test, write how many isopods you think will choose each side of the tray. If you are using ten isopods, don't forget your number for each test should add up to ten.

MY PREDICTION:

TEST ONE

WARM	COOL

TEST TWO

WET	DRY

TEST THREE

SHADE	LIGHT

Now test your isopods. Remember to always be gentle to your live subjects. For each test, write in the number of isopods that ended up on each side.

TEST RESULTS:

TEST ONE

WARM	COOL

TEST TWO

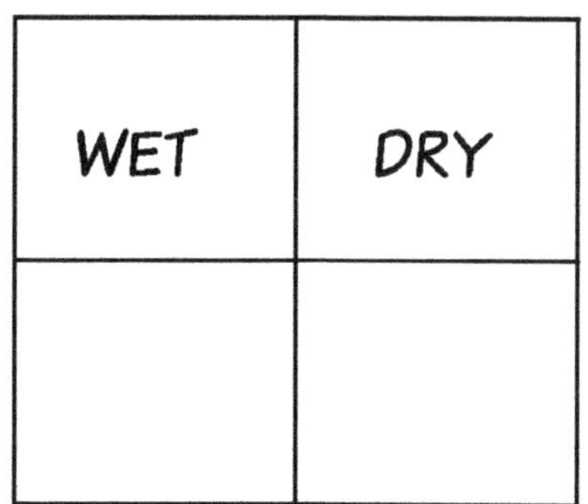

WET	DRY

TEST THREE

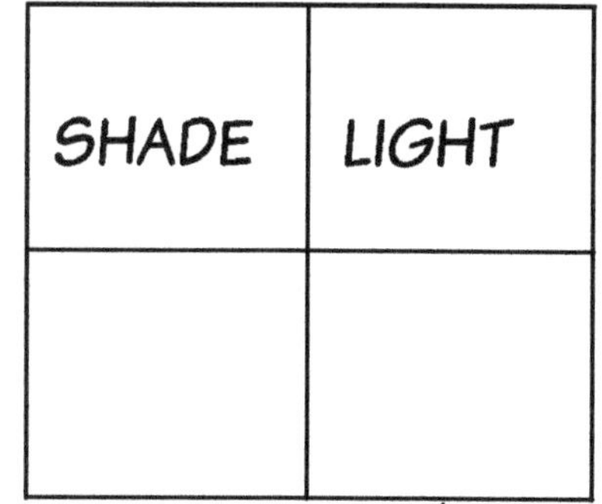

SHADE	LIGHT

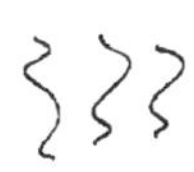

Crustacean Lab #2: A HOME FOR ISOPOD- pg. 2

Fill in each graph below to show the results of your isopod experiments.
Color in one rectangle for each isopod that chose that side.

TEST ONE

WARM										
COOL										

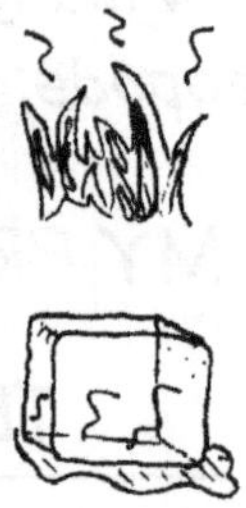

TEST TWO

WET										
DRY										

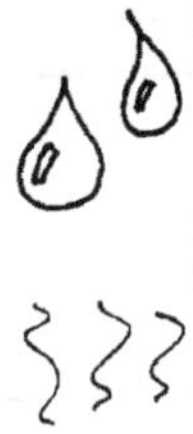

TEST THREE

SHADE										
LIGHT										

I think isopods will stay moist if they are in areas that are____________

____

Most of my isopods went to the sides that were__________________

____.

Most of my isopods (circle one) 1. Make good choices 2. Are in deep trouble

ANIMAL KINGDOM BOOK: ARTHROPOD - instructions

Material:

copy of lab sheet (1 page), pencil

Animal Kingdom Book (assembled)

scissors

glue gel (Elmer's School Glue Gel)

art supplies- markers, colored pencils, stickers, rubber stamps etc.

The Usborne Illustrated Encyclopedia of The Natural World or other animal encyclopedia (optional)

Procedure: Lab Day:

1. Use the information from your science notebook to fill in sections of your lab paper. List the characteristics of arthropods, examples of arthropods and in the trivia section write the one thing about them that you thought was the most unusual or interesting. You may want to use the internet or *The Usborne Illustrated Encyclopedia of The Natural World* to find a really neat fact about them.
2. Color the animals and the large arthropod label on the page.
3. Page 5 in your Animal Kingdom Book is for your arthropod summary. Cut out the small arthropod label (number 1). Have your book open to page 1 (cnidaria) The arthropod label will go on the small tab for page 5 that shows at the top and to the right of the echinoderm label. Glue your arthropod la bel in place.
4. Turn to page 5. Cut out all remaining sections of your lab paper. Arrange them on page 5 of your book in whatever order you enjoy. Glue them in place.
5. Finish decorating this page as you wish.
6. Use your book to review what you have learned about the animal kingdom.

Possible Answers:

1. Characteristics of arthropods:
   - jointed legs
   - exoskeleton
   - segmented body
2. Examples of arthropods:
   - butterfly (insect)
   - spider (arachnid)
   - lobster (crustacean)
   - fly (insect)
   - tick (arachnid)
   - crab (crustacean)

NAME ______________________________ DATE ___________________

ANIMAL KINGDOM BOOK- ARTHROPOD

1. ARTHROPOD

2. 

CHARACTERISTICS OF ARTHROPODS

____

____

____

EXAMPLES OF ARTHROPODS

____

____

____

____

____

TICK

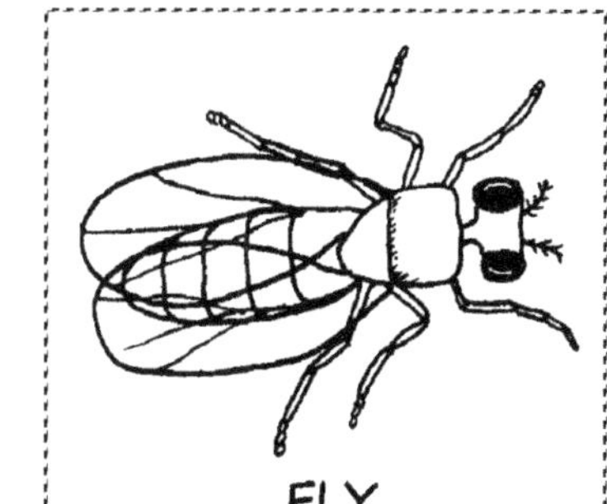

FLY

INCREDIBLE ARTHROPOD TRIVIA

____

____

____

____

____

____

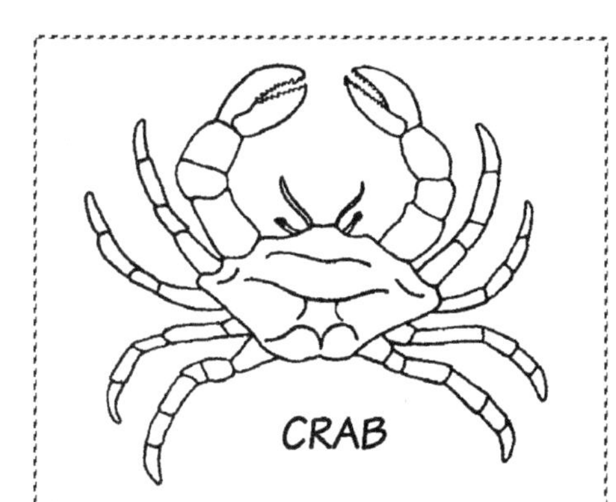

CRAB

NAME ______________________ DATE ______________

For my notebook

VERTEBRATES HAVE BONES INSIDE

Awhile ago we talked about how your spinal cord runs down your back inside a stack of donut shaped bones. These bones are called vertebrae. There is a whole string of them that runs from the bottom of your skull to the tip of where your tail would start, if you had one. Feel from your skull down your back and you can feel this string of bones. Vertebrate animals are animals that have vertebrae, or backbones. So far we have studied only invertebrates which are animals with no backbones, or any other bones. Many of the animals we have learned about have soft, squishy or see-through bodies. It's easy to tell that they have no bones. Can you remember some of these? How about worms, slugs or sea jellies? Remember that some mollusks have a shell and arthropods have an exoskeleton but none of them have bones. We are going to learn about vertebrate animals now.

There are 5 major groups of vertebrates. They are fish, amphibians, reptiles, birds and mammals. When most people talk about animals, they usually only think about vertebrates but now you know there are many animals like worms and sea stars that have no backbones. As you learn about vertebrates, see if you can guess which one of the 5 groups you belong to.

Vertebrate Activity: INVERTEBRATE OR VERTEBRATE? - instructions

Material:

1 large piece of construction paper or poster board
marker scissors
magazines with animal pictures (invertebrates too!)
glue gel (like Elmer's School glue gel)

Aloud: You have studied many invertebrate or boneless animals. Today you will separate animal pictures into two groups—one for invertebrates or animals without backbones and one for vertebrates or animals with backbones. As you decide which group each animal belongs to glue it onto the proper side of your paper to display all the different kinds of animals you have learned about.

Procedure:

Lab Day:

1. Cut out small pictures of animals from magazines. Don't forget to include invertebrates such as insects, mollusks (snails and their kin), sea stars, worms, etc.
2. Hold construction paper horizontally and draw a line down the middle from top to bottom. On one side write "Invertebrates" and on the other side write "Vertebrates."
3. Determine what is an invertebrate (anything they have studied so far) and what is a vertebrate (only fish, amphibians, reptiles, birds and mammals).
4. Glue animals on the appropriate side of the page.

Conclusion/Discussion:

1. Review the purposes of bones. Without bones to hold them upright, invertebrates must either be fairly small or live in water so the water can hold them up. Can you think of any tall, boneless animals that live on the land? There aren't any.

For More Lab Practice:

1. Label each invertebrate on the poster with the phylum it is in.
2. Research: Compare the tallest land vertebrate with the tallest marine invertebrate. How tall is the tallest marine invertebrate?
3. Start an invertebrate zoo! Using critter keepers, jars, plastic shoe boxes etc. learn how to collect, house and feed various invertebrate animals. Never keep an animal you can't properly care for. For information on keeping various critters alive read *Pets in a Jar* by Seymore Simon. This is a wonderful book that tells you how to find, house and keep water striders, planaria (flat worms), brine shrimp, earthworms, crickets and many more.

NAME ______________________________ DATE ____________________

For my notebook

FISH BREATHE THROUGH GILLS

Do you have a fish at home? Take a look at it (or one in a book). What makes a fish a fish? How is it different from a cat or a snake? First look at its body. A fish's body is covered with scales, not hair or feathers. Watch closely and see if you can see it breath. Do you see the slit on each side of its head? These are flaps that cover the very delicate gills underneath. Fish don't breathe with lungs like we do. Watch him gulp water into his mouth and then open his gill covers so the water can leave. Water passes over his gills and they take the oxygen out of the water. Even though they live in water, fish breathe oxygen, just like you do. Now, look where his arms and legs should be. You see fins, don't you? Arms and legs don't work as well for swimming as fins do. So, you can see fish are covered with scales, breathe through gills and have fins. Some things you probably can't see are that they lay eggs and are cold blooded. Cold blooded means that their bodies change temperature as the water does. You can take your fish's temperature just by putting a thermometer into his water. Wouldn't that be funny if you went to the doctor and he took your temperature just by putting the thermometer on the table next to you? This doesn't work for people because we are warm blooded. No matter how hot or cold the air is our body temperature doesn't change very much. Now, let's see how your fish measures up.

Fish Lab #1: MEASURE A FISH - instructions

Material:

copy of lab sheets (2 pages), pencil
1 live fish- (big goldfish works well) in bowl or tank with water
1 small, clear plastic jar (just big enough for fish to be in)
gram scale or triple beam balance (for weighing fish)
small fish net
1 room thermometer (for measuring temperature of room and fish's water)
1 sterilized human thermometer (for measuring body temp. of child)
1 watch or clock with second hand
1 centimeter ruler

Aloud: We read about the things that make a fish a fish. Now we are going to measure, count parts of and weigh your fish. We are even going to take your fish's temperature without touching him. Let's see how many characteristics of a fish we can see today.

Procedure:

Lab Day:

1. Observe the fish, then draw in box with as much accuracy as possible.
2. For question #3, time your child for one minute while he watches his fish "blink." This is sort of a trick question because, except sharks, fish have no eyelids, hence no blinking.
3. For question #5 time your child for 10 seconds as he counts the times his fish opens its mouth to breathe. Multiply this answer by 6 to find out how many times in one minute the fish takes a breath.
4. For Part II: Measuring; do your best to measure the fish as it swims by. A round bowl is less accurate than a tank as it enlarges the fish but this is all about practice, not accuracy. For # 9 subtract the fish's body length from the whole length (body + tail fin - body) to find the length of the tail fin.
5. For Part III: Weighing, fill small plastic jar with enough water from fish's bowl to cover the fish once he's in. Weigh the jar and water and record on line 11. Gently net fish and put him in the jar. Weigh again and record on line 10. Now, subtract weight of jar and water from weight of jar, water and fish and you will be left with the weight of the fish.
6. For Part IV shake down any thermometers that don't go down on their own (such as body thermometers). Place room thermometer into fish bowl to find water temperature. The fish's body temp. will be the same as this reading. Now take the temperature of the room and then take the temperature of the child WITH THE STERILE THERMOMETER.
7. Fill in the graph to show how the temp. of a cold blooded animal is the same as the temp of it's environment but our body temp. isn't. Hopefully the room isn't body temp.!

Conclusion:

1. Discuss results of lab. How is it better to be cold blooded? How is it worse?
2. What kinds of things do warm blooded animals do to stay warm when the weather gets cold?
3. Imagine going out in the snow if you were cold blooded. What would you do differently?

For More Lab Fun:

1. Go to different locations where it is hot, cold, medium. Take the air temp. and your body temp. Graph the results and discuss whether or not body temp. changed with each location.

NAME ______________________ DATE ______________

FISH LAB #1: MEASURE A FISH - pg. 1

Look closely at your fish and draw it in the box. Draw in some of his scales.

Label your fish's head, body, tail, mouth, eye, gill cover and 3 of his fins.

Part I: Counting

1. My fish has __________ fins (how many?)
2. My fish has ___________ gill covers.
3. My fish blinks _________ times in one minute.
4. My fish has _____________ eyelids.
5. By watching his mouth I see he breathes (A)_______ times in 10 seconds
6. (A)______ X 6 = _______. My fish breathes _______ times a minute.

Part II: Measuring: My Fish's Size

7. My fish is about (B)______ centimeters (cm) long from nose to end of tail.
8. My fish's body (without fins) is about (C)_______ cm. long.

9. (B) - (C) = length of tail (B)______ cm - (C)______ cm = ______ cm

   My fish's tail is ___________ cm. long

Part III: Weighing: My Fish's Weight:

DO QUESTION 11 BEFORE YOU DO QUESTION 10! (see instructions)

10. Weight of Jar + Water + Fish ____________ grams (g)
11. minus - Weight of Jar + Water ____________ g

12. equals = Weight of Fish ____________

    My fish weighs ___________ g.

FISH LAB #1: MEASURE A FISH - pg. 2

Part IV: My Fish's Temperature

13. My fish's water temperature is ________° F
14. My fish's body temperature is the same as his water's temp. because he is cold blooded. This means his body temp. is about ________° F
15. The room temperature is ________° F
16. My body temp. is ________° F. I am warm blooded. This means my body temperature can be different than the room temperature.

	water temp. °F	fish's temp. °F		room temp. °F	my body temp. °F
105°					
100°					
95°					
90°					
85°					
80°					
75°					
70°					
65°					
60°					
55°					

By studying my fish I could actually see that fish: (check all that you could <u>see</u>)

- ☐ have scales
- ☐ have fins
- ☐ breathe with gills
- ☐ are cold blooded
- ☐ lay eggs
- ☐ breathe in the water

Fish Lab #2: FISH FLOATERS - instructions

Material:

copy of lab sheet (1 page), pencil
clear plastic drink bottle with cap
glass eye dropper (big hunky plastic ones won't work)
water
colored pencils

Aloud: Birds are build perfectly for flying, even on the inside. Fish are build perfectly for swimming, even on the inside. Have you ever gone swimming with air filled floaties before? Can you swim down in the water with them? Would you have to work hard to stay underwater? Fish can't work hard to stay at each depth. It would use all of their energy. They must be able to adjust the amount of air inside their bodies, like we do with floaties. Fish have a swim bladder inside that they can fill up with air or empty out as they swim. As they get down to the depth they want, they adjust the amount of oxygen in these bladders until they float where they want to. We are going to do an experiment to show how the amount of air can make an object float at different depths.

Procedure:

Lab Day:

1. Fill the bottle with water.
2. Drop the empty dropper into the bottle. Where does it float?
3. Fill the dropper with water to the very top of the black bulb part. You will have to shake water down into the black bulb and then squeeze and fill again in order to get it all full. Drop it into the bottle. Where does it float? Empty the dropper again.
4. Now, fill the dropper to right where the glass meets the rubber top. Drop it into the bottle. Fill the bottle with water until it is bulging over. Put the cap on. Where does it go now? Now squeeze the bottle very hard. What happens with the dropper? (it should drop down in the bottle)
5. Squeeze the bottle and hold it so that the "fish" will stay half way down in the bottle. What you are doing is adjusting the amount of air in the dropper, with pressure. This happens to fish as they dive up and down in the water. By adding or removing air from their swim bladder, they can stay at any given depth without working at it.

Possible Answers:

#2. air or oxygen
#3. float near the top
#4. float near the bottom

For More Lab Fun:

1. Good swimmers can experiment with going underwater after taking a small breath and then after a large breath. Which way was it easier to go to the bottom? Can you dive to the bottom with floaties on? Can you see how adjusting the amount of air in (and on) your body can make you float at a different depth?

For Your Information:

This is a classic experiment showing how air pressure and volume are related. Your children will see this again in later years and will learn to better understand what is making the dropper go up and down. For now, letting them imagine a fish under water is fine. Because it looks like a magic trick, hopefully when they are asked to do the same experiment they will say "I always wondered how that worked."

Fish Lab #3: FISH FLOATERS Instructions

[illegible]

[illegible] Birds are [illegible] perfectly for [illegible] the inside. Fish [illegible] perfectly for swimming, even on the inside. Have you ever done swimming [illegible] filled floaties [illegible] down in the water [illegible] them? Would you have to work hard to stay [illegible] can't work hard to stay at each depth. It would use all of their energy. They need [illegible] adjust the amount of air inside their bodies like we do with floaties. Fish have [illegible] that they can fill up with air or empty out as they swim. As they get down to the depth they want, they adjust the amount of oxygen in these bladders until they float where they want. We are going to do an experiment to show how the amount of air can make an object float at different depths.

[illegible]

[illegible] Answers

[illegible]

[illegible] Float [illegible]

[illegible]

For [illegible]

[illegible] to see how changing the amount of [illegible] make you float at a different depth.

[illegible] Information

[illegible] your children [illegible] see this [illegible] and will learn to better understand [illegible] the dropper go up and down. [illegible] hopefully when they [illegible] to do the same [illegible] wondered how that worked.

NAME ______________________________ DATE ______________________________

Fish Lab #2: FISH FLOATERS

Fish have a swim bladder that they can fill with or empty of air in order to float at different depths.

1. Using a different colored pencil for each one, draw in each bottle where your eye dropper fish floated.

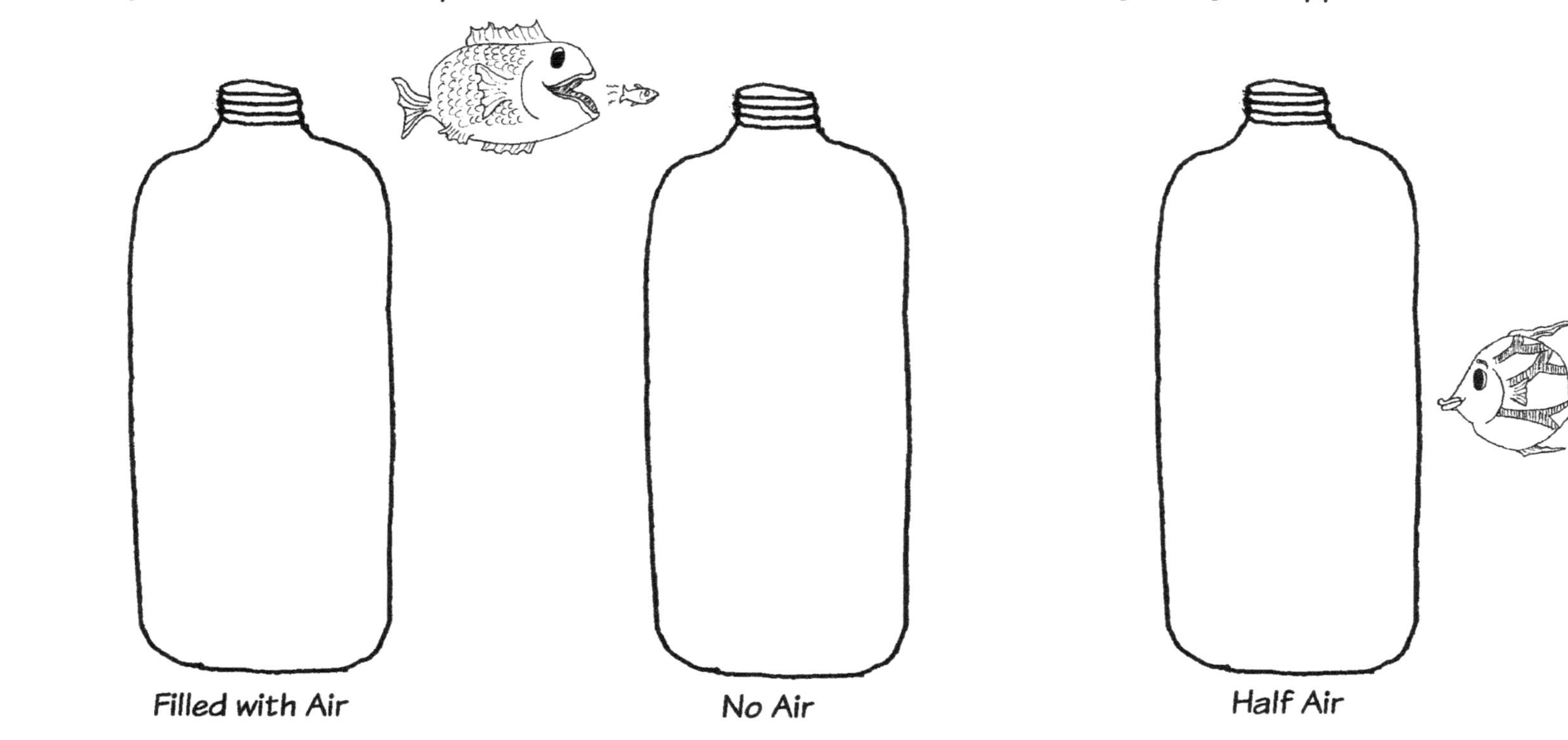

2. To swim at different depths, a fish changes the amount of ____________ in its body.

3. With a lot of air, a fish will ____.

4. With just a little air in her bladder, a fish will ______________________________.

NAME ______________________________ DATE ____________________

For my notebook

AMPHIBIANS LEAD A DOUBLE LIFE

Did you know that toads live two different lives? Toads hatch from eggs underwater. They spend the first part of their lives breathing through gills and swimming around like fish. The adult stage of their lives they spend hopping around on land and breathing with lungs, just like yours. Toads, along with frogs and salamanders are amphibians. The word "amphibian" means "double life." Remember when an insect, like a butterfly, changes from one form to another it is called metamorphosis? Amphibians go through metamorphosis too. They hatch from eggs laid in water and start life with a tail, gills and no legs. After some time, legs appear, lungs develop and, in frogs and toads, the tail disappears. Even though adult amphibians are ready for a life on land they have to stay near water. Their skin is very thin. In fact, they breath through their skin also. If their skin gets dry, water will leave their bodies through the skin and they will die. Amphibians are also cold blooded like fish. In cold areas you will not hear frogs in the winter because they are dormant or sleeping. They cannot keep themselves warm like birds and people can.

So, baby amphibians are like fish. They hatch from eggs, have no legs, swim in water, breath through gills and are cold blooded. Adult amphibians are very different from fish. They have legs, breath through lungs and their skin and live mostly on land. Amphibians sure do lead a "double life," don't they?

Amphibian Lab #1: METAMORPHOSIS WHEEL - instructions

Material:

copy of metamorphosis pattern sheet (1 page), pencil
2 manila file folders
scissors
metal fastener (brad)
gel glue (like Elmer's School Glue Gel)
colored pencils

Aloud: Now you know that butterflies aren't the only animals that go through metamorphosis. Frogs, toads and salamanders do too! There is a difference though. Butterflies go through 4 stages and amphibians only go through 3. Today you are going to make a wheel to show the differences between complete, four-stage metamorphosis, like a butterfly does, and incomplete, three stage metamorphosis, like an amphibian does.

Procedure:

Lab Day:

1. *Have kids color the 7 critter diagrams on page 227. This page has several patterns on it and you will use them one at a time. Don't cut without knowing where you are going!*
2. *First, cut out the big, main square – the one with everything else inside of it. Place on fold of one file folder, as indicated. Draw around square. Cut through both sides of the folder, leaving the fold uncut.*
3. *On one side of the other folder, draw around and cut out the circle pattern.*
4. *Now, from the file you have cut into a square, cut out the section as indicated, opposite the fold. This is your viewing window, once you put your decorated circle inside the square. Cut through both layers of folder.*
5. *Cut out your labels "Complete Metamorphosis" and "Incomplete Metamorphosis" and glue one onto each side of the square, near the fold.*
6. *Place the pattern back on the square folder. Mark where the center hole will go. Place the folder circle inside the square. Center it and punch through the hole, through all three layers. Connect together with the brad.*
7. *Cut out all 7 animal pictures.*
8. *For this step be cautious: If you use too much glue, your wheel won't turn. Overgluers should wait until their glue dries before turning the wheel and going to the next step! Turn to the side that says "Complete metamorphosis." Glue the butterfly egg in the viewing winder, onto the circle. Write "Egg" in the window under the egg. Turn the circle 1/4 turn and glue in the caterpillar. Write "Larva" under the caterpillar. Turn 1/4 turn again and repeat with chrysalis. Label it "Pupa." Finally glue down the butterfly and label it "Adult"*
9. *Turn the wheel over and do the same thing with the incomplete metamorphosis, this time spacing them 1/3 turn apart. Label the egg "Egg," the tadpole "Tadpole" and the frog "Adult."*
10. *Use your Metamorphosis Wheel to learn about and review the stages of complete and incomplete metamorphosis*

PATTERNS FOR METAMORPHOSIS WHEEL

COMPLETE METAMORPHOSIS

INCOMPLETE METAMORPHOSIS

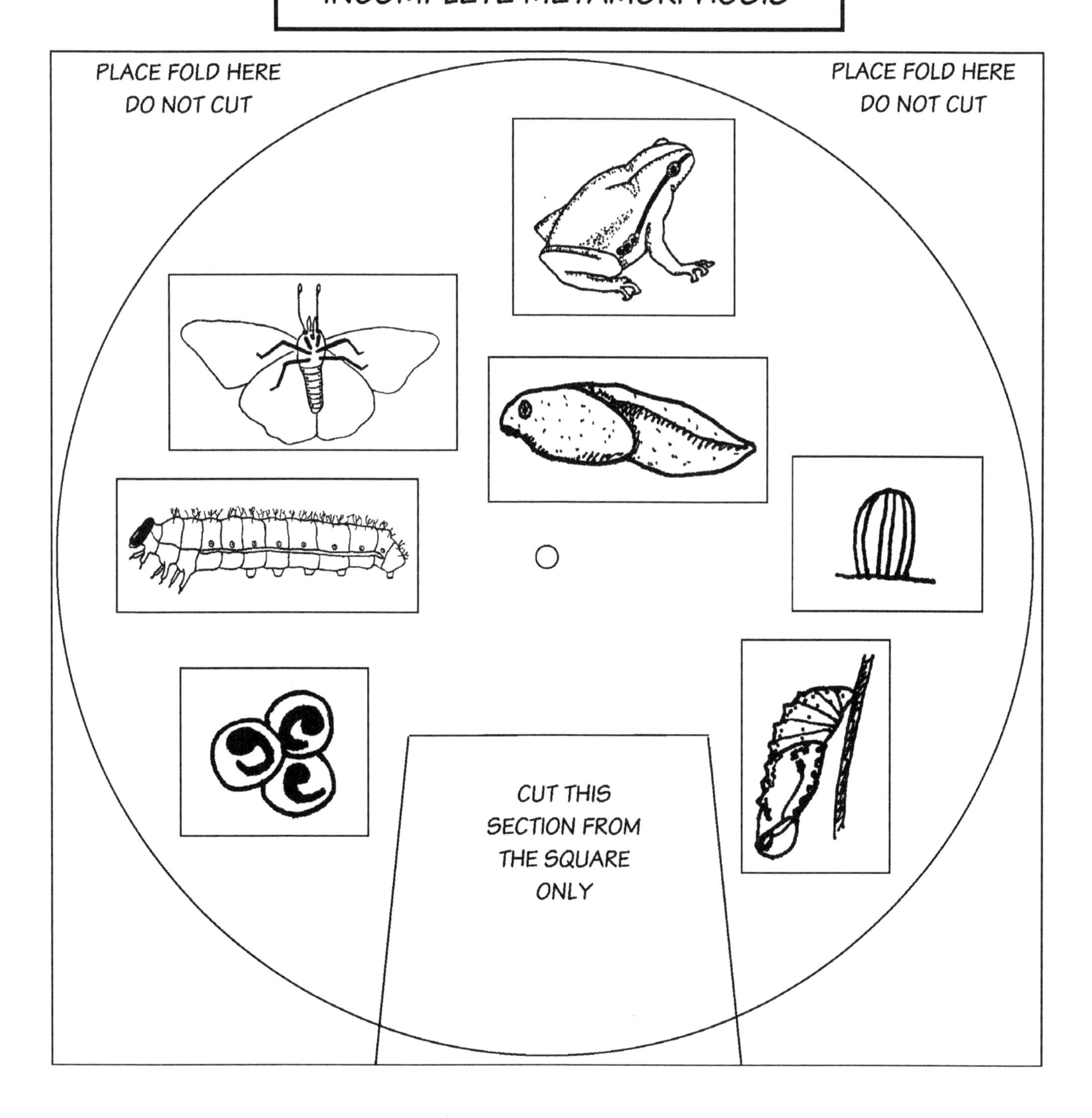

Amphibian Lab #2: NO "EAR" IN HEARING - instructions

Material:

copy of lab sheet (1 page), pencil
metal Slinky toy
metal pie pan
metal spoon
salt
eye dropper
water table

Aloud: Have you ever noticed that not all animals have ears like we have? You might think that these animals can't hear, but you would be wrong. Fish "hear" with a line that runs along each side of their bodies. Newts and salamanders "hear" with their legs! Hard to believe until you understand that hearing is just sensing movement. Movement causes waves in the water, ground or air. When these waves go through the air they hit our ears and we hear. The same happens when waves in the water hit a fish or waves in the ground reach a salamanders legs. We are going to do several experiments to show how these waves move.

Procedure:

Lab Day:

1. *This works best on a smooth surface, like a wood floor. To show how a wave moves through a substance, have a different person hold each end of the Slinky. The person on one end should hold their end still while the other person quickly pushes their end towards the other a few inches, compressing the coils. Do this in a quick, forward jerk. You should see the Slinky coils compress in a wave all the way to the other end of the Slinky. This shows how waves push particles (molecules) along from one area to another.*
2. *Have two people hold the Slinky again but this time one person moves their end rapidly in a wave motion side to side. You should see a wave move along the Slinky.*
3. *Place the pie pan upside down on a table. With a thin layer of salt, write a "Z" across the upturned bottom of the pan. Tap the pan with the spoon several times and watch the salt grains.*
4. *Now turn the pan over and add about 1/2" of water. Drop one drop of water right in the middle of the water. Watch how the waves move through the water.*
5. *Have somebody on the other side of a table tap the table softly enough that they can barely hear it. Now try with your ear on the top of the table. Was it easier to hear sound through the air or through the wood of the table? Place your ear on the ground (wood or tile) and see if you can hear somebody walking across the floor. Is it louder through the air or through the floor?*
6. *Complete the lab sheet.*

Possible Answers:

3. *on the table (sound travels better through wood than through air)*
4. *Sound travels faster through the ground so you can hear animals coming from farther away with your ear to the ground - like a salamander hears, except he uses his legs. No need for him to bend down.*

Conclusion / Discussion:

1. *When you tap on a fish's tank it disturbs them. Is it hurting their ears?*
2. *After doing the experiments, do you think people hear better than a salamander would?*

For More Lab Fun:

1. *Try to make a sound that doesn't include any movement. (It's impossible. Sound IS movement).*

Amphibian Lab #3: NO "EAR"ly HEARING – Instructor

[illegible]

[illegible] that [illegible] you would [illegible] a line that runs along [illegible] understand that [illegible] movement [illegible] waves in the water [illegible] When these waves go through [illegible] The same [illegible] in the water hit a fish or waves in the ground [illegible]. We're doing [illegible] experiments to show how [illegible] waves move.

[illegible]

[illegible] the lab sheet.

Possible answers:

[illegible]

[illegible]

For More Lab Fun:

[illegible]

NAME ______________________________ DATE ____________________

Amphibian Lab #2: No "Ear" in Hearing

To SEE sound waves:

1. Tap a pan with salt on it to see how the waves move the salt

This is what my salt pattern looked like before I tapped the pan:

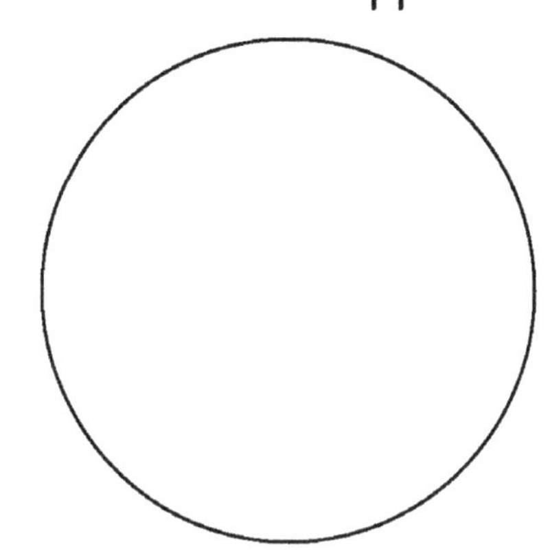

This is what my salt pattern looked like after I tapped the pan:

2. Draw the waves that were made when the drop landed in the middle of the pan.

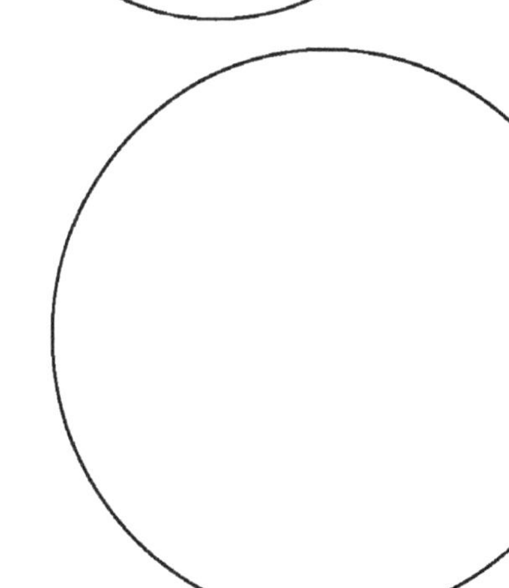

To HEAR sound waves:

3. The table tapping was louder when my ear was ____________________

4. In old cowboy movies they put their ear on the ground to hear bison or horses coming. Why do you think they did that?

____

NAME ______________________________ DATE ___________________

For my notebook

REPTILES HAVE SCALY SKIN

Alligators, turtles and lizards, Oh my! What do these three animals have in common? They are all reptiles and so are crocodiles and snakes. In fact, dinosaurs were reptiles too. In the days of the dinosaurs, reptiles ruled the planet. Some reptiles were over 30 meters long. Measure how long that is with a meter stick. What! No gigantic amphibians hopping the earth? What is it that reptiles have that amphibians don't?

For one thing, reptiles have bodies covered with hard plates or scales. This covering keeps their body's water in so reptiles don't have to stay wet to stay alive. Reptile eggs are very different too. Amphibian eggs are soft and unprotected but reptiles lay eggs with a tough, leathery shell. With a shell like this, reptile eggs can be laid on land. When baby reptiles hatch they look like tiny adults. They don't go through metamorphosis. Another advantage that reptiles have for living on land is their claws. Amphibians have no claws but reptiles do and they can use them for climbing, fighting and digging.

Reptiles are cold blooded, like fish and amphibians but with their tough, dry scales, their claws and their strong egg shells, they can do many things and live in many places amphibians can't.

Reptile Lab: HOW DRY I AM - instructions

Material:

copy of lab sheet (1 page), pencil
cookie sheet
waterproof marker
3 paper towels
water
zipper-type plastic baggie

Aloud: Amphibians breathe through their skin. Along with the air, they also lose water through the holes in their skin. Because of this, they must stay very near water or they will dry out. Reptiles have tough, scaly skin to prevent this water loss. They do not breathe through their skin so they don't lose water through their skin. How about humans? Do we lose water through our skin? Yes, we do—when we sweat. We don't need to keep our skin soaked like an amphibian does, but we do need to drink water regularly to prevent dehydration. This lab will show how water loss compares between amphibian, reptile and human skin.

Procedure:

Lab Day:

1. With a waterproof marker, label one paper towel "Amphibian #1." Label the second paper towel "Human #2" and label the third paper towel "Reptile #3."
2. Wet paper towel #1 until it is soaked, but not dripping. Lay it flat on one end of the cookie sheet.
3. Wet paper towel #2, roll it up and lay it on the sheet next to #1.
4. Wet paper towel #3, roll it up and then seal it in the baggie. Place it on the cookie sheet also.
5. Answer question #1 on lab sheet.
6. Leave the cookie sheet in the sun until the first sheet is completely dry. This could take anywhere from 30 minutes to several hours, depending on the weather, so check often.
7. Record on lab sheet which towel dried out first. Leave the setup in the sun until the second paper towel is dry. Complete lab sheet.

Possible Answers:

#2. First = #1 Second = #2 Third = #3
#3. Frog or amphibian Because it dries out quickly or because it lets out the water
#4 Person or Human Because it loses water but stays wet on the inside or loses water more slowly than an amphibian
#5 Lizard or Reptile Because it keeps water in.
#6 Lizard
#7 Drink a lot of water
#8 They stay near water.

Conclusion / Discussion:

1. What are the advantages of being a reptile that can travel away from water?
2. What are the advantages of being human and being able to re-hydrate (satisfy our water needs) by drinking as opposed to having to keep our skin wet all the time?

Reptile Lab: HOW DRY AM I? Instructions

[illegible]

NAME ______________________ DATE ______________

Reptile Lab: HOW DRY I AM

HYPOTHESIS: (MY BEST GUESS):

1. I think the paper towels will dry out in this order:

First- # _____ Second- # _____ Third- # ______

TEST (THE EXPERIMENT):

2. The paper towels dried out in this order:

First- #______ Second- #_____ Third- # _____

CONCLUSION (WHAT I LEARNED)

3. The flat paper towel is like the skin of ____________________

because ______________________________________

____

4. The rolled paper towel is like the skin of____________________

because ______________________________________

5. The wrapped paper towel is like the skin of ____________________

because ______________________________________

6. Which animal would do best in the desert? lizard frog human

7. People have holes (pores) in their skin to let out heat. This also lets out water. How can we compensate for this water loss on a hot day?

____

8. How do amphibians keep from drying out?____________________

____

NAME ______________________ DATE ______________

Reptile Activity: LIZARD POEM

THE LIZARD

The lizard is a timid thing
That cannot dance or fly or sing;
He hunts for bugs beneath the floor
And longs to be a dinosaur

- John Gardner

Illustrate this poem in the box below

NAME ______________________________ DATE ___________________

For my notebook

BIRDS HAVE FEATHERS

How can you tell a bird when you see one? Birds all have one thing that no other animals have. Can you think of what that is? Birds all have feathers. Birds are also warm blooded, which means their bodies, like yours, don't change temperature with the weather. Whether it is hot or cold outside their temperature doesn't change. Did you remember that you can take a fish's temperature by putting a thermometer into his water? Invertebrates, fish, amphibians and reptiles are all cold blooded but only birds and mammals are warm blooded. In fact, a bird's feathers aren't just for flying. They also help him to stay the same temperature.

Birds have two legs but they also have two wings for flying. Some birds can't fly but they still have wings. Some, like penguins, use their wings to "fly" through the water!

Birds also lay eggs with a hard, brittle shell, unlike the flexible, leathery shells of the reptiles' eggs or the gooey, unshelled eggs of the amphibians.

So, we know that birds have feathers, are warm blooded, have wings and lay brittle eggs. They also all have beaks that tell a lot about how they live. Look in a book or at the zoo at the variety of beaks that birds have. No wonder so many people enjoy bird watching. It seems no two beaks are alike.

Bird Lab #1: BIRDS OF A FEATHER - instructions

Material:

copy of lab sheets (2 pages), pencil
zoo, wildlife center or other place where a variety of birds are available to see and have their names posted
binoculars (optional)

Aloud: There are thousands of different types of birds, all surviving in different ways. Some are smaller than your hand, some taller than your parents. Some have delicate beaks for sipping nectar, some have thick beaks to crack open nuts. Scientists have grouped birds based on how they survive and where they live. Today we are going to see a number of different birds and try to put them into the groups that scientists have chosen for them.

Procedure:

Lab Day:

1. As you wander through the zoo, study each bird you see, even the wild ones, if they let you! Try to place each bird into its appropriate group on the lab sheet. To simplify things, not all groups are included, therefore, help children do their best but remember, the point is not to "get it right" as much as to understand the process. Help guide the children to notice characteristics but let the answers be their own.
2. Fill in the lab sheet as indicated.

Conclusion / Discussion:

1. Why are the birds at the zoo different from those around home? (Because somebody has hand selected an interesting variety of birds from around the world for you to see. Usually they select the ones that make us "Ooh" and "Ahh" the most.)
2. Why are the birds at the beach different from the ones in a park or on the prairie? (The environment dictates what the bird must look like. An ostrich would not do well climbing trees to find insects, like a woodpecker does, but an ostrich does just fine running in vast areas avoiding large predators. Everything about a bird is perfectly suited to where the bird lives and how it survives—its size, feet, beak, body shape, feathers, neck length, wing size etc.)

For More Lab Fun:

1. While on your field trip take photos of the birds you see. Start an album of the bird groups you run across.
2. Choose a migratory bird to study further. Write a travel log for your bird. Include maps of migration routes.

NAME ______________________ DATE ______________

Bird Lab #1: BIRDS OF A FEATHER- pg. 1

Go on a bird hunt! Which of these kinds of birds can you find? Write them on the lines below.

FLIGHTLESS BIRDS
(ostriches, penguins)

____
____
____
____

HEAVY, GROUND DWELLING BIRDS
(turkeys, quail, grouse)

____
____
____
____

SHORE AND WATER BIRDS
(ducks, herons, pelicans, gulls)

____
____
____
____

BIRDS OF PREY
(owls, hawks, vultures)

____
____
____
____

PERCHING BIRDS
(sparrows, crows, pigeons)

____
____
____
____

THICK BILLED, SEED CRACKING BIRDS
(parrots, parakeets)

____
____
____
____

OTHER BIRDS
(hummingbirds, woodpeckers)

______________________ ______________________
______________________ ______________________
______________________ ______________________
______________________ ______________________

Bird Lab #1: BIRDS OF A FEATHER- pg. 2

Make a graph of how many birds you found from each group.

Flightless										
Ground dwelling										
Shore and water										
Birds of Prey										
Perching										
Thick billed										
Other										
	1	2	3	4	5	6	7	8	9	10

At the zoo we saw more ____________________ than any other group.

In my neighborhood I see more ____________________

At the beach I would expect to see more ____________________

In the boxes below draw a...

bird of prey beak	seed cracking beak	any water bird's beak

Bird Lab #2: HOW DUCKS STAY DRY - instructions

Material:

2 small paper sacks
sink or bowl to work over
vegetable oil (about 2 Tbl.)
water
mister (optional)

Aloud: All birds have feathers but no other animals do. Feathers have many uses. We know that feathers are very light weight and help the bird fly. They are also used to keep the bird warm. Have you ever seen a bird on a wire on a cold day, all puffed up and fat looking? They fluff out their feathers to trap warm air inside and this keeps their bodies warm. Without feathers, water birds, like ducks, would probably freeze to death in the winter. Swimming in cold weather is as deadly for birds as for people. So, how do ducks stay warm? Well, they must first stay dry. To do this they spend a lot of time "preening" or fixing, fluffing and oiling their feathers. Birds have an oil gland on their backs at the base of their tail. They spread the oil from this gland all over their feathers many times a day. Because oil and water are pushed away from each other, the oil in the feathers keeps the water out and the duck dry! Remarkable, huh? Let's see how this happens.

Procedure:

Lab Day:

1. Use your fingers to spread a little oil onto one side of one of the paper sacks. Do not get oil on the second sack. Wash your hands with warm water and soap.
2. Over a sink or bowl, spray (or drip) water onto the plain sack. What does the water do? Now spray water on the oiled sack. What does it do?

Conclusion / Discussion:

1. How does the water react differently on the two sacks? (Water on the oiled sack beads up and runs right off. Water on the plain sack doesn't bead up and some of it soaks in).
2. How does oil keep a bird dry? (The same way it keeps the bag dry- by repelling water).
3. People who swim in cold water have been known to spread some kind of grease or lard over their bodies. Why do you think this is? (It keeps them drier and warmer).
4. Not all birds swim. How does spreading oil on their feathers help the birds that don't swim? (Keeps them dry and warm in a cold rain).
5. How would flying be more difficult with wet feathers? (Wet feathers would be heavier and not maintain the proper shape for aerodynamics to occur).

For More Lab Fun:

1. If you have birds of any kind, with permission, mist them with water. Many birds really like it. Watch what the water does when it hits their feathers.
2. Do this lab with real bird feathers instead of paper sacks.

Bird Lab #3: ARE YOU MY BROTHER? - instructions

Material:

copy of lab sheets (3 pages), pencil
scissors
colored pencils

Aloud: After looking at real birds at the zoo and seeing how scientists have put them into groups based on their feet, beaks and body shapes, you have become a local bird expert. Explorers have found an island with some unknown and unusual birds on it. Your job is to organize these birds into 3 or 4 groups by finding ways in which they are alike. Choose only one characteristic that you want to use to group your birds by.

Procedure:

Lab Day:

1. Cut along lines on lab pg. #3 to separate bird pictures.
2. Lay pictures in random order.
3. Children group them as they wish. Make sure they have decided whether to group by feet, beaks or body shape. They can only use one characteristic. There are no right or wrong answers as long as there is a reason for the groupings.
4. If children have trouble getting started, suggest one characteristic such as type of feet. Move all birds with long toes into one area. To make things easier, the birds have been drawn with 3 different types of feet, 3 types of bodies and 4 beak types. Any of these characteristics (or characteristics of their own discovery!) will work to help them with groups. Allow children to take it from there.
5. In the first box on lab pg. #1, have kids right the numbers of all birds from one of their groups. Write why those birds have been placed together. Continue filling in lab page for all the remaining groups.

Possible answers:

1. (parrot) crushing or opening seeds
2. (duck) swimming or walking in mud
3. (eagle) catching prey, fish, food . . .
4. (heron) spearing food, fish, frogs . . .
5. (owl) camouflage, hiding . . .
6. (flamingo) walking in water

9 - 14 Try to help kids come up with at least four of these on their own. Possible answers are: lay eggs, have wings, walk on 2 feet, have feathers, have beaks, are warm blooded, lay brittle eggs.

For more lab practice:

1. Regroup your birds based on different characteristics.
2. Cut bird pictures from magazines and organize them based on the groupings you have made, or make new groups with the magazine pictures.
3. Go outside with your lab paper. Try to fit the wild birds from your neighborhood into the groupings you have made.
4. Get your child his own bird field guide (suggestions available under "Book Suggestions"). Look through it together and then go outside to see what birds you see there. Binoculars are pretty necessary also, but you can get started without them if you must. Start a bird watching "life list" of the birds you see everywhere you go. There is a blank life list provided after this lab. This makes for a wonderful family hobby that you can learn to do together. Let your child place a tiny sticker in his field guide next to each bird he identifies in the wild.

Bird Lab #3: ARE YOU MY BROTHER? - Instructions

NAME ______________________________ DATE ____________________

Bird Lab #3: ARE YOU MY BROTHER?- pg. 1

I grouped my birds like this:

GROUP 1 — These birds are all alike because ______________________________

____

GROUP 2 — These birds are all alike because ______________________________

____

GROUP 3 — These birds are all alike because ______________________________

____

GROUP 4 — These birds are all alike because ______________________________

____

These are special features that some birds have. Name a possible purpose for each feature.

1. Thick, crushing beak (parrot) ______________________________
2. Webbed toes (duck) ______________________________
3. Long, sharp talons (toenails) (eagle) ______________________________
4. Long, spearlike beak (heron) ______________________________
5. Bark colored feathers (owl) ______________________________
6. Very long legs (flamingo) ______________________________

Bird Lab #3: ARE YOU MY BROTHER?- pg. 2

Draw your own bird in the box below. BE CREATIVE!

7. My bird is called a ______________________________

8. He belongs in group number ______ because he ______________

____

I have learned that, because of where they live and what they eat, birds are very different from each other but I also know that all birds are alike in many ways. I know that all birds...

9. ______________________________

10. ______________________________

11. ______________________________

12. ______________________________

13. ______________________________

14. ______________________________

Bird Lab #3: ARE YOU MY BROTHER? - pg 3

Copy and cut out the birds for use during Bird Lab #3

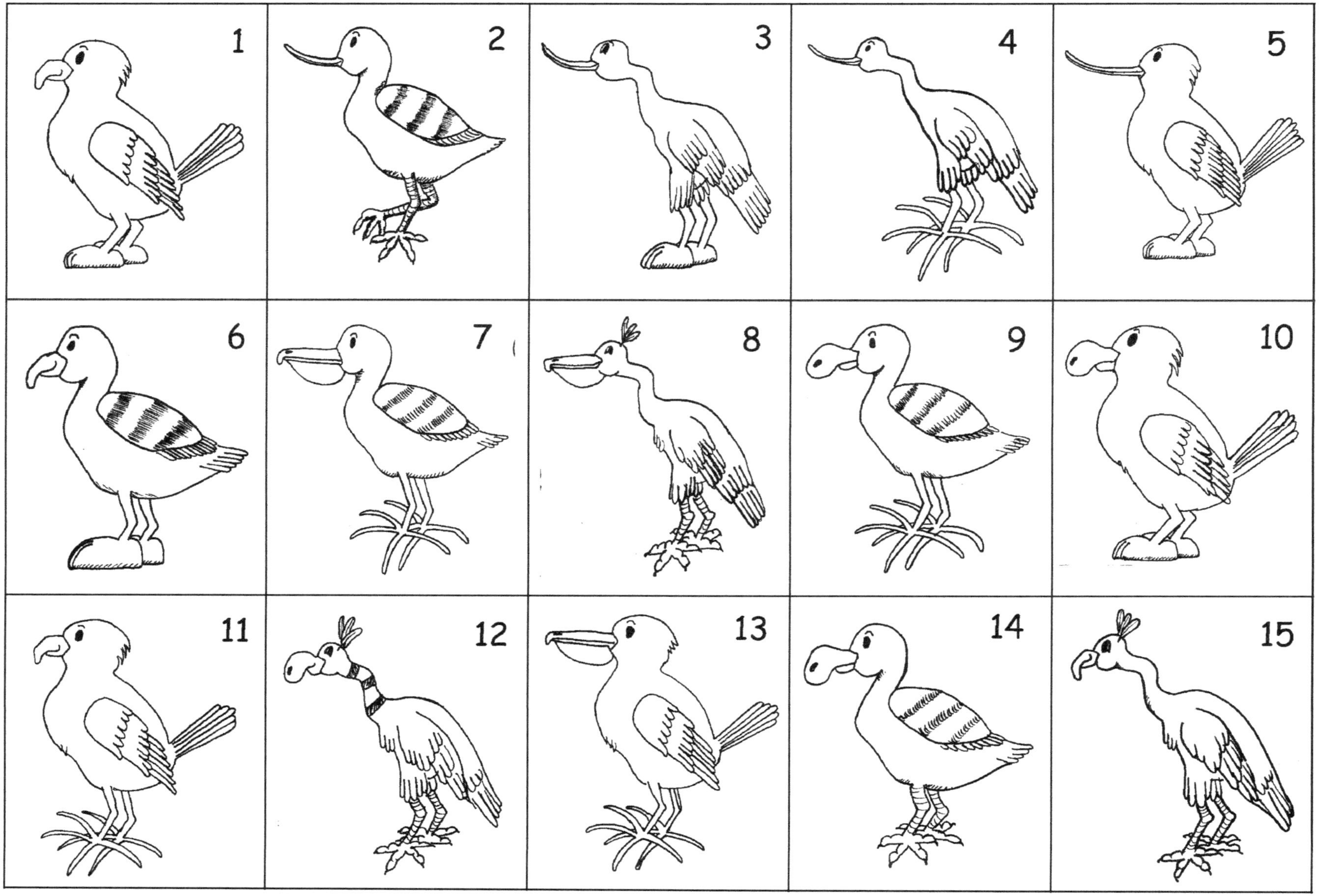

1	2	3	4	5
6	7	8	9	10
11	12	13	14	15

NAME Joe Birder

MY WILD BIRD LIFE LIST

DATE	SPECIES (type of bird)	**	HABITAT	LOCATION	NOTES
7/12/2011	Bald Eagle	4	lake	Millerton Lake at Finegold Creek	immature, no white band
8/14/2011	House sparrow	5	city	Sacramento- 16th & F street	eating seed at bird feeder
8/14/2011	Mallard duck	4	pond	Polly Park, Dana, Virginia	may have been tame

** is for degree of certainty, 1-5: 1 = just guessing 5 = positive

NAME ____________________ MY WILD BIRD LIFE LIST

DATE	SPECIES (type of bird)	**	HABITAT	LOCATION	NOTES

** is for degree of certainty, 1-5: 1 = just guessing 5 = positive

NAME ______________________________ DATE ____________________

For my notebook

MAMMALS DRINK MILK

For a long time you have known that you are not a fish or bird. You can't breath underwater and you aren't covered with feathers. You also know you aren't an amphibian or reptile. You didn't hatch from an egg nor are you cold blooded. So what type of vertebrate are you? You are a mammal. The word "mammal" comes from the latin word "mamma," meaning breast. What all mammals have in common is that they nurse their young. In other words, all baby mammals grow up on their mother's milk. Do birds or fish? No, only mammals. Even mammals that live underwater, like dolphins and whales feed their babies milk.

Like birds, all mammals are warm blooded. Remember, warm blooded animals stay the same temperature inside no matter what the temperature of the air is. This means their bodies must have ways of staying warm in cold weather and ways of staying cool in hot weather. Birds have feathers to keep warm, so what do mammals have? Fur or hair. A bald fox or polar bear would never live through the winter. In fact, a polar bear's hair is actually a hollow tube that draws heat down to the bear's black skin. It not only keeps the heat in, it pulls it in!

So, from the tiniest mouse to the tallest giraffe, all mammals like you are raised on milk, are warm blooded and have hair on their bodies. These very simple things have allowed mammals to live from hot deserts to frozen ice.

Mammal Lab #1: WHAT'S IN A NAME? - instructions

Material:

copy of lab sheets (2 pages), pencil
clipboard
zoo (if you have no zoo nearby you can do this in a field or forest but you'll probably want to take along field guides to help you with the names of the animals you see)

Aloud: You know what a bat is but did you know there are over 40 kinds of bats in the United States? How do you know which one you've seen? Well, many animal names are very helpful in identifying the animal you are looking at. I bet if I put 5 different bats in front of you, you would know which one was the Spotted Bat, even if you've never seen one before. Would you know a Gray Fox from a Red Fox if you saw them both together? Probably, because the names tell what they look like. Good common names help us to identify an animal. Another helpful type of name tells us where an animals lives or where it comes from. Do you think you'll be likely to see an Arctic Fox in Texas? How about a River Otter in the ocean? Today we are going to the zoo to find as many animals as we can with good descriptive names. While you're there, you can earn mega-bonus points for spotting a mammal doing what only mammals can do-nursing their young!

Procedure:

Lab Day:

1. As you wander through the zoo, write the names of the animals with descriptive and/or place names. If you are lucky enough to have a well equipped zoo near you, you may want to limit the activity to only mammals. Also write down any mammals you spot nursing.
2. Write the points in the appropriate columns. If the animal is a mammal, give yourself 5 extra points.
3. Discuss how the descriptive name helps identify that animal. Is it a good, accurate name? Some can be misleading. Black Bears come in black, brown, "blue," reddish and even nearly white.
4. Complete the second lab paper.

Answers:

#3. Bonus question- Lewis's Woodpecker was named after Meriwether Lewis of the Lewis and Clark expedition. He wrote about seeing them near Helena, Montana. A Lewis's Woodpecker skin is the last known actual specimen left from the expedition.

For More Lab Fun:

1. Spend another day at the zoo on a "spot and stripe hunt." Find out how different patterns help these animals survive.
2. Another good zoo hunt if for camouflaged animals. Try to find the Walking Stick or the Screech Owl. Describe how their coloration protects them. Then, use sticks, cones and leaves to build your own "camouflaged critter."
3. Find other animals that have been named after people. Research who they are named for and why.

NAME ______________________ DATE ______________

Mammal Lab #1: WHAT'S IN A NAME?- pg. 1

To score:

"LOOKS" names = 10 points (Yellowbelly, Spotted, Great...)

"PLACE" names = 10 points (Northern, Desert, Mexican...)

20 points if the name has both (Alaskan Brown Bear, Mountain Cottontail...)

5 points extra if the animal is a mammal

50 bonus points for every mammal you see nursing - no matter what its name is!

	(10)	(10)	(5 extra)	(50)
ANIMAL'S NAME	LOOKS	PLACE	MAMMAL	NURSING

Mammal Lab #1: WHAT'S IN A NAME?- pg. 2

To add up your scores from the first page, count by 10's to total your score in the LOOKS and PLACE columns. Count by 5's down the MAMMAL column and count by 50's to total your score in the NURSING column. Add these together to find your naturalist score.

COLUMN TOTALS:

LOOKS = ____________

\+ PLACE = ____________

\+ MAMMAL = ____________

\+ NURSING = ____________

= TOTAL SCORE = ____________

Your naturalist rating:

0 - 100 points = Fledgling naturalist- Good start!
105 - 150 = Apprentice naturalist - Keep up the good work!
155 - 200 = Accomplished naturalist - Your hard work has paid off!
205 and above = Master Naturalist - You are an inspiration to all!

How do good, descriptive names help us identify animals new to us?

____

____

Some animals are named after the people who first described them. Are these names helpful in identifying the animal?

____

Bonus question:
Who is the Lewis' Woodpecker named for? ____________________

Mammal Lab #2: COAT OF BLUBBER - instructions

Material:

copy of lab sheet (1 page), pencil
small bowl or wide mouthed cup
ice water
2 identical, small plastic bags (sandwich size or smaller)
2 room thermometers (science type) to test cold water
(body temp. thermometers don't work for this)
1/2 cup of lard or shortening
colored pencils- red and blue

Aloud: Do you remember which two groups of animals are warm blooded? Birds and mammals, right? Well, that means that birds and mammals must have special ways to stay warm, even in the bitter cold. Many mammals have special hair to keep themselves warm and other mammals have a thick layer of fat or blubber to keep out the cold. Since water takes the warmth right out of you, marine mammals, like whales and seals have an extra thick layer of blubber. Did you know there is a bear that is classified as a marine mammal? Can you guess which bear it is? It lives most of its life in and ON the water- even when it is frozen. The polar bear is a marine mammal, just like a dolphin. We are going to do an experiment today to show how blubber keeps the body warm.

Procedure:

Lab Day:

1. Put ice water into bowl to about 3" deep.
2. Place shortening into a bottom corner of one baggie. Push one thermometer into the center of the shortening. Shape a thick layer of the shortening around the thermometer. Put the second thermometer into the other baggie with no shortening.
3. Record the temperature from each thermometer. Put the bulb end of each thermometer (still in the baggie) into the bowl. After one minute, record the temperature on the chart again. Record again each minute for the next 5 minutes.
4. Use the data you gathered to do the graph. Use a different colored pencil for each thermometer so the lines are easier to follow. This is the first time the children have done a line graph so they may need some help plotting the points on the graph.

Conclusion:

1. Which temperature dropped more quickly? Which temperature dropped farther? Did the shortening work well as an insulator?
2. If animals that live in cold climates need to keep heat in and animals that live in hot climates must find ways to give off heat, which would be better off with a thick layer of blubber? What would happen to a hot weather animal with a thick blubber layer?
3. Can you think of any animals that are thin during hot weather and put on a thick layer of fat for the winter? What other changes do animals go through to get ready for winter (thicker layer of fur, some even change color to match winter snow)

For More Lab Fun:

1. Do the same experiment but instead test your fingers. How long can you keep a finger in the ice water without insulation and how long with? Record the times. It's best to do this test one at a time. Use the uninsulated finger first, then test with a finger with shortening molded around it.

NAME ______________________________ DATE ____________________

Mammal Lab #2: COAT OF BLUBBER

PREDICTION:

I think the shortening will make the temperature ____________________

TEST:

Check the temperature on both thermometers at the times indicated and record on the chart below.

	START	1 MIN.	2 MIN.	3 MIN.	4 MIN.	5 MIN.	6 MIN.
BLUBBER							
NO BLUBBER							

Plot your results on the graph below. Draw a red line to connect the "blubber" dots and a blue line to connect the "No blubber" dots.

A polar bear's blubber can be 4 ½ inches thick.

A whale's can be up to 16 inches thick!

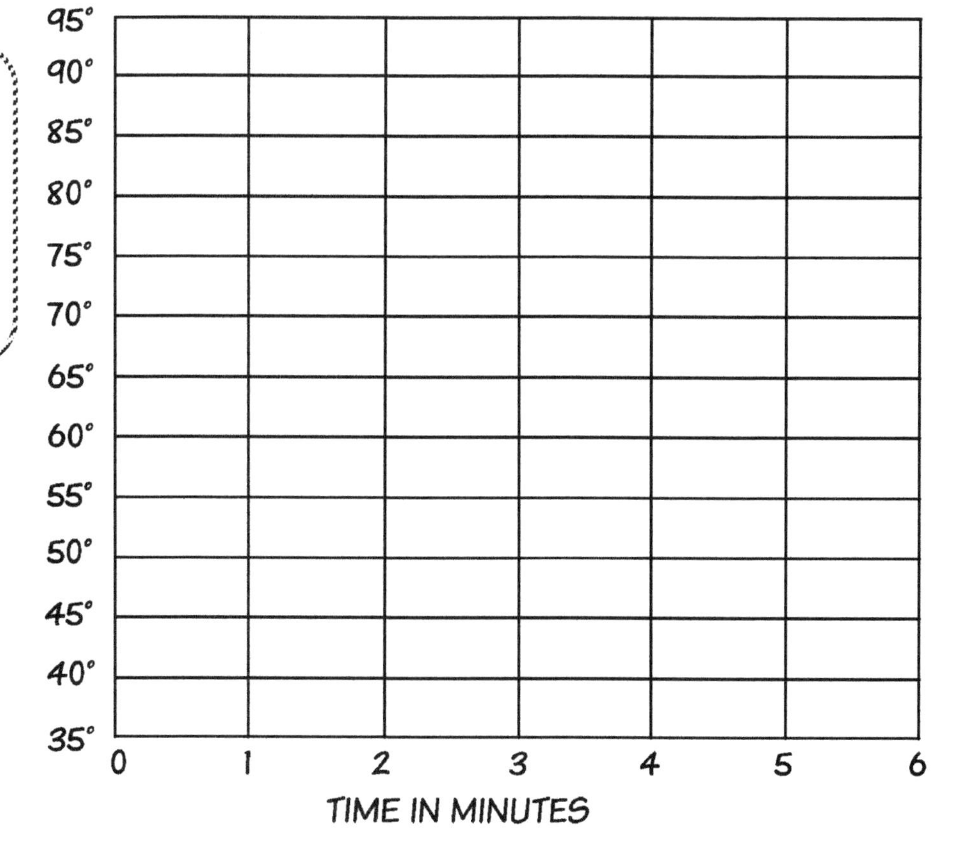

ANIMAL KINGDOM BOOK: VERTEBRATE - instructions

Material:

copy of lab sheet (1 page), pencil

Animal Kingdom Book (assembled)

scissors

glue gel (Elmer's School Glue Gel)

art supplies- markers, colored pencils, stickers, rubber stamps etc.

The Usborne Illustrated Encyclopedia of The Natural World or other animal encyclopedia (optional)

Aloud: This is the last page you will put into your Animal Kingdom Book. You should have a page for cnidaria, worms, mollusks, echinoderms, arthropods and now vertebrates. There are other groups in the animal kingdom, but these are the major ones. You will study some others in a few years. You should start to have a sense of what kinds of animals are in each group. Refer to your Animal Kingdom Book often and these animals will start to become as familiar as your own relatives!

Procedure:

Lab Day:

1. Use the information from your science notebook to fill in sections of your lab paper. List the characteristics of vertebrates, examples of vertebrates and in the trivia section write the one thing about them that you thought was the most unusual or interesting. In the "examples" section, try to include one example from each group of vertebrates (one for fish, one for amphibian etc.) You may want to use the internet or *The Usborne Illustrated Encyclopedia of The Natural World* to find a really neat fact about them.
2. Color the animals and the large vertebrate label on the page.
3. Page 6 in your Animal Kingdom Book is for your vertebrate summary. Cut out the small vertebrate label (number 1). Have your book open to page 1 (cnidaria). The vertebrate label will go on the small tab for page 6 that shows at the top and to the right of the arthropod label. Glue your vertebrate label in place.
4. Turn to page 6. Cut out all remaining sections of your lab paper. Arrange them on page 6 of your book in whatever order you enjoy. Glue them in place.
5. Finish decorating this page as you wish.
6. Share your finished book with relatives. Show them what you have learned about the animal kingdom. Use your book to review what you have learned about the animal kingdom.

Possible Answers:

1. Characteristics of vertebrates:
   - bones
   - spinal cord
   - complex organs
2. Examples of vertebrates:
   - goldfish (fish)
   - frog (amphibian)
   - snake (reptile)
   - flamingo (bird)
   - human (mammal)

NAME ______________________ DATE ______________

ANIMAL KINGDOM BOOK- VERTEBRATE

1. VERTEBRATE

CHARACTERISTICS OF VERTEBRATES

EXAMPLES OF VERTEBRATES

FROG

2. VERTEBRATE

AMAZING VERTEBRATE TRIVIA

SNAKE

FISH

Animal Kingdom Summary: WHAT AM I? - instructions

Material:

copy of Animal Summary- Animal Samples (page 1) for each child
copy of Animal Summary- Phylum Names (page 2)
copy of Animal Summary- Phylum Characteristics and hints (page 3) (optional)
scissors
glue gel (Elmer's School Glue Gel)
large sheet of butcher paper (approx. 3' x 6')
samples of critters, 4-5 from each animal phylum studied is a good number- shells, dried sea star, rubber or plastic invertebrate toys, live spider or insects in small vials. Use your imagination! The more "real" things you have the more fun it is.

Aloud: You have learned about a lot of different animals. Today we are going to try to remember how our animal kingdom is organized. You will get a number of different animals and your job will be to put them into the groups where you think they belong. Do your best and have fun!

Procedure:

Prior to Lab: Parent/Teacher

1. With as much secrecy as possible, gather a variety of vertebrate and invertebrate animals, pictures, sea shells, etc. Try to have 4-5 representatives from each group the children have studied. You can include small stuffed animals, a plastic lizard, rubber snake, real clam shell, vial with real spider or insect or isopod (roly-polies), gummy worm, teddy bear counter, anything goes. The more 3-D variety the more the kids like it. Put them all into a basket or bin to hand to each child.
2. The Animal Summary- lab page 1 provides 2-3 samples from each animal phylum, in case you have trouble getting samples for a group. Cut these out to add to the basket.
3. Lay butcher paper onto the table or floor and draw lines vertically to divide paper into 6 even sections. At the top of section 1 glue the title "Cnidaria," section 2 - "Worms," section 3 -"Mollusks," section 4 - "Echinoderms," section 5 - "Arthropods," and section 6 - "Vertebrates."

Lab Day:

1. Invite the kids in. Before handing out the samples, brainstorm together the major characteristics of each group. Write (or glue from the page provided) a list of 2-3 characteristics for each group under each heading. Use the Phylum Characteristics page for help. You can also draw, have the kids draw or glue samples from the Phylum Characteristics page of basic body plan to help the non-reading students.
2. Hand out baskets of samples and let the kids place each sample where they think it belongs on the page.

Conclusion Discussion:

- Go through the kids choices. Discuss what they put where. If animals are in the wrong spots, give clues to help them come up with the right answers. Move animals as needed.

For More Lab Fun:

- Take a picture of your kids in front of their animal phylum summary. This is too hard to store but the picture will last forever as a reference.

For Your Information:

The phylum for vertebrates is actually called "Chordata." It includes some very simple, sea jelly looking animals with a nerve cord but no bones. We feel this is too much information for children this age so we have borrowed the subphylum name "vertebrate" and used it for the phylum. If you prefer, and think your children can handle it, you might choose to use the proper term of "Chordate."

Animal Kingdom Summary: WHAT AM I?- page 1

Copy and cut out the animals for use during Animal Kingdom Summary

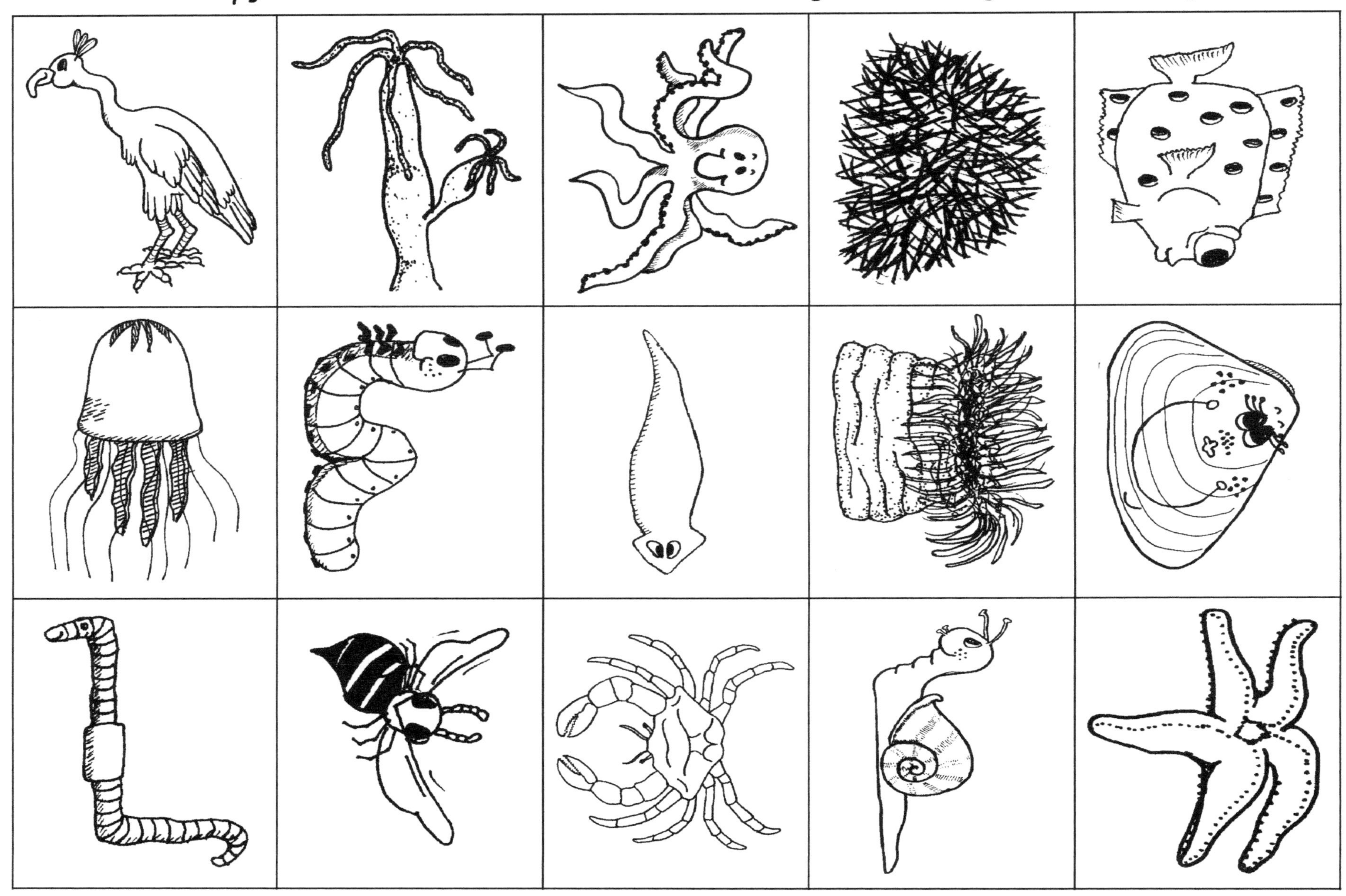

CNIDARIA

WORMS

MOLLUSKS

ECHINODERMS

ARTHROPODS

VERTEBRATES

Animal Kingdom Summary: Phylum Characteristics - page 3

Cut out and glue to butcher paper or use for reference

CNIDARIA

(sea jellies, coral, sea anemones)

- Hollow body
- Stinging tentacles
- No organs

WORMS

(not caterpillars)

- No legs
- Long body
- A head and a tail side
- Round, flat or segmented

MOLLUSKS

(snails, slugs, octopi, clams, squid, limpets)

- Soft body
- Usually with hard shell
- Organs inside

ECHINODERMS

(sea stars, sea urchins, sand dollars, sea cucumbers)

- Spiny skin
- Body in five parts
- Tube feet for moving
- All live in water

ARTHROPODS

(insects, spiders, crabs, lobsters, shrimp, centipedes, roly-polies)

- Jointed legs
- Six or more legs
- Hard exoskeleton
- Three or more body parts
- Many with wings
- Many with antennae

VERTEBRATES

- Fish, amphibians, reptiles, birds and mammals
- Bones
- Spinal cord

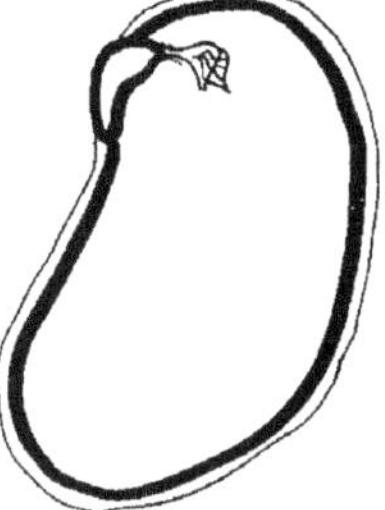

THE PLANT KINGDOM

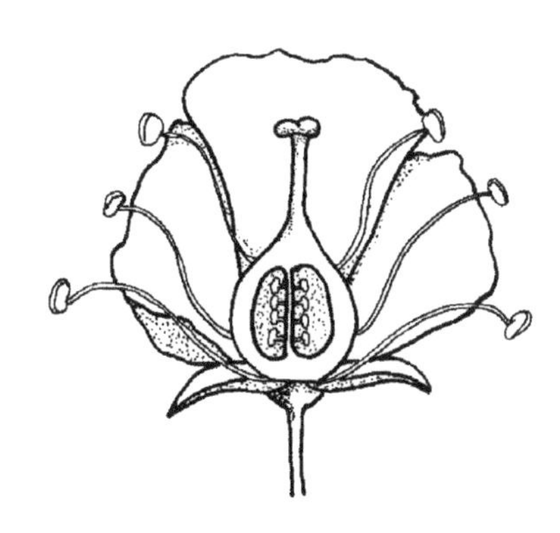

NAME ______________________________ DATE ___________________

For my notebook

FLOWERS MAKE SEEDS

I love flowers, don't you? You can plant a garden or walk through a field or even a desert and see beautiful plants but going to one of these places when it is full of flowers is even more wonderful. Flowers are great to look at, to smell, to draw and just to sit quietly among. It seems like plants make beautiful flowers just to make us happy. Well, a flower's job is to be attractive, but not to us. They want to look and smell good to insects like bees, beetles and even flies and to hummingbirds, bats and other animals. So, why would plants want all these odd visitors? To help them reproduce. The male part of a flower makes pollen which must be taken to the female part of a flower so it can make seeds. Seeds, of course make more plants. Animal visitors like bees pollinate flowers by carrying pollen from one flower to another. A flower that cannot be seen or smelled will attract no pollinators and must rely on wind or water to carry its pollen for it. Some flowers even have "targets" for the insects to follow. Look at a pansy. It has a different color right in the middle where the insects need to go to find the pollen. Can you find any more that have this? Some flowers even smell like a female bee so that the male bee will stop in for a visit. Next time you look at a flower, think about how it is attracting its pollinators.

Flower Lab #1: COLOR THE FLOWER - instructions

Material:

copy of lab sheet (1 page), pencil
crayons or colored pencils: blue, orange, green, pink and brown
marker or crayon: black

Aloud: In order for seeds to form, a flower must be pollinated. Sometimes the wind blows the pollen onto the female reproductive parts, but usually insects, birds or bats pollinate flowers. When a bee crawls into a flower the pollen sticks to its abdomen and is carried to another flower. If the pollen touches the top of the pistil, it sticks. Each pollen grain grows a tube down through the pistil to where the ovules are. One sperm cell will leave each pollen grain, travel down the pollen tube and enter an ovule. One sperm cell is needed for each ovule. Now the ovule is fertilized and can grow into a seed.

Procedure:

Lab Day:

1. Color the flower as indicated.
2. Make sure your child understands where the pollen comes from and what it must stick to. You might also want him to use the marker to draw a pollen tube growing down to the ovary. (See diagram below.)

Possible answers:

1. This flower could make 8 seeds. Count the ovules.
2. Many answers would be fine. Examples: bee, hummingbird, bat, beetle . . .

Conclusion / Discussion:

1. Show me where a bee needs to touch to pick up the pollen. Now where will the pollen go? (top of the pistil)
2. Show me where the pollen tube will grow to and where the sperm cell needs to go (down to the ovule).
3. Why do flowers have such pretty petals? Why do flowers smell sweet? (to attract pollinators)
4. If a sweet flower attracts bees and hummingbirds, what do you think a flower smells like if it is pollinated by flies? (some flowers that are pollinated by flies look and smell like rotting flesh)
5. What do you think a flower might look and smell like if it is pollinated by the wind? (Many flowers are pollinated by wind. They are typically very small, green, not showy at all and have no distinct odor. There is no reason for them to be showy, smelly or attractive.)

For Your Information:

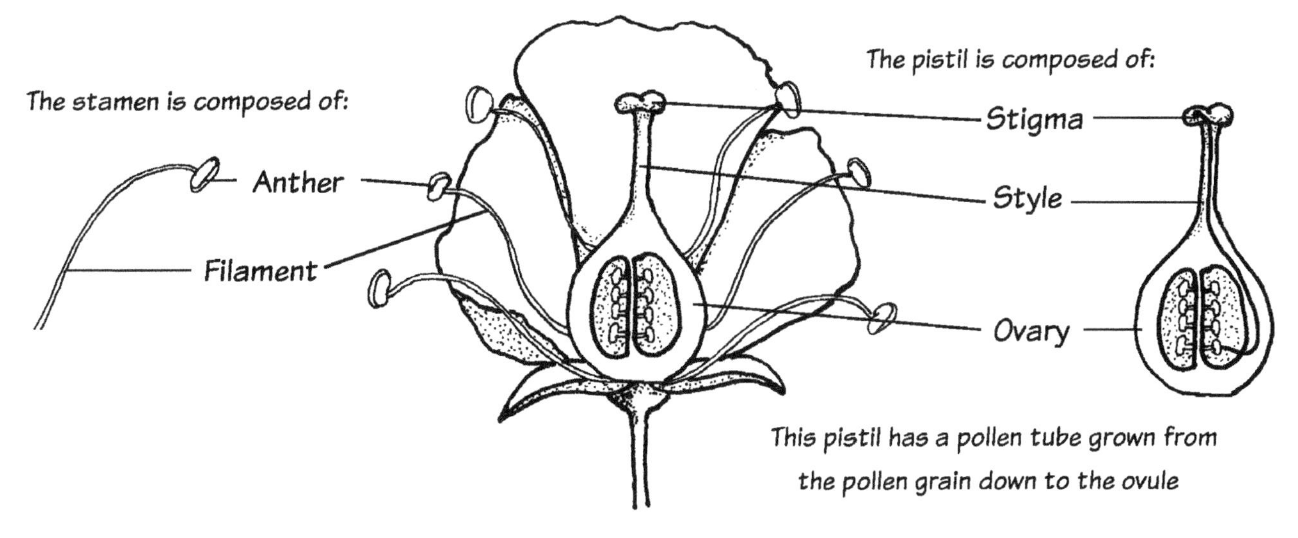

This pistil has a pollen tube grown from the pollen grain down to the ovule

NAME ______________________ DATE ______________

Flower Lab #1: COLOR THE FLOWER

JOB	NAME	COLOR
Female reproductive parts	Pistil	orange
Male reproductive parts	Stamen	blue
Covers the flower before it opens	Sepals	green
Colorful to attract pollinators	Petals	pink
If fertilized, will become seeds	Ovules	brown

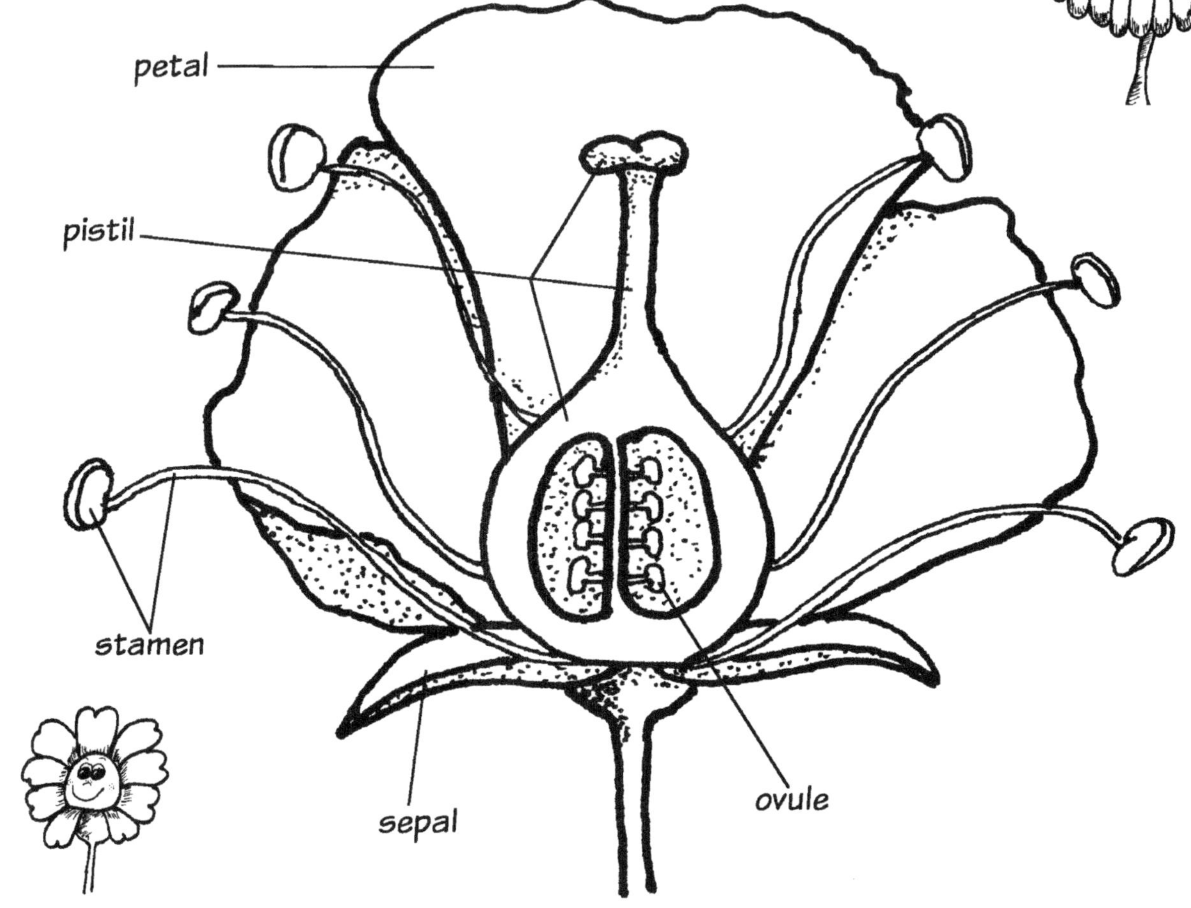

For seeds to be formed, pollen must get from the top of a stamen to the top of the pistil. With a marker, draw a line to show the path of the pollen from a stamen to the pistil.

1. How many seeds could this flower make? ______________

2. What type of animal might pollinate a flower like this? ______________

Flower Lab #2: WHAT MAKES UP A FLOWER - instructions

Material:

copy of lab sheet (1 page), pencil
completed Flower Lab #1: Color the Flower (for reference only)
one flower- gladiolus is best, petunia good also although you need to tell the children that the petals are fused together, some lilies and azaleas work too. The thing to look for is a flower big enough to see the parts. It should have a few, well defined sepals, a good pistil and a reasonable number of petals and stamens. Some flowers have dozens of petals and stamens and are not good for learning on.
clear tape
centimeter ruler
tweezers (optional)
hand lens / magnifying glass (optional)

Aloud: When you dissect something you cut it open and see what it's made of. Today you are going to dissect a flower and try to find and identify all of those parts you heard about in the last lab. Remember, just like every person, every flower is different so just do your best and have fun getting up close and personal with your flower. Try to use as many senses as possible (except taste of course). Look at, smell and feel your flower's parts.

Procedure:

Lab Day: *DON'T PASS OUT THE FLOWERS YET!*

1. Review the parts of the flower from the coloring page. Before looking at your flower guess how many parts you will find. Write down your guesses.
2. Count and separate the parts, taping one of each in the spaces provided. Fill in the lab sheet as indicated.

Conclusion / Discussion:

1. Review the purposes of the different flower parts.
2. With the flower pulled apart, show the path of a pollen grain, from the stamen to the top of the pistil and down the pistil into the ovary (bulb part at the bottom of the pistil).
3. Discuss what makes your flower attractive to pollinators. Would it be able to reproduce if you were a bee? In other words, would you be attracted to this flower?

For More Lab Fun:

1. Pick some wild flowers. Count the parts of the wildflowers and compare to your "store bought" flower.
2. Start a wild flower collection. Press wild flowers (or the entire plant, when it isn't too big) between sheets of paper. Place heavy books or bricks over them or use a real plant press. Make sure to note type of plant (if you can), where it was found and date collected. There are many more flowers than birds so they are harder to identify, but some field guides are suggested in the Suggested Reading list under "Field Guides." After they are dry, mount the plants in a notebook and label with the information. NOTE: Some wildflowers are rare and some are protected by law. Before picking, make sure your flowers are legal to pick and abundant.
3. Press some flowers (as described above) and use them to make cards for friends and relatives.

NAME ______________________________ DATE ____________________

Flower Lab #2: WHAT MAKES UP A FLOWER?

Get out the flower coloring sheet you did in Flower Lab #1. Review each of the parts of the flower and then guess how many of each you think you might find on the flower you will be looking at today. **No fair peeking until you have written down your guesses!**

FLOWER PART	MY GUESS	MY COUNT
Sepals		
Petals		
Pistils		
Stamens		

This is how my flower attracts its pollinators:

My petals are colored ____

My flower smells____

My flower has these special features ____________________________

____

If I were a bee or hummingbird I would go to my flower because __________

____

Tape one petal here My petal is __________cm. long	Tape one stamen here My stamen is __________cm. long
Tape one sepal here My sepal is __________cm. long	Tape one pistil here My pistil is __________cm. long

NAME ______________________________ DATE ___________________

For my notebook

SEEDS GROW INTO PLANTS

Can you name some plant seeds that you eat? How about peas, rice, coconut or popcorn? A large number of the foods we eat are seeds. Look around your kitchen and see how many different seeds you can find. A seed is a special package that contains a tiny baby plant and food to help it start growing so it can become a big plant. There are two main types of flowering plants, dicots (**die**-cots) and monocots (**moe**-noe-cots). Dicots have seeds that split into two parts. "Di" means two and "cots" is short for cotyledons (cot-uh-**lee**-dunz)—the seed parts. Have you ever eaten peanuts and noticed how they fall into two pieces? Each half is a cotyledon. Monocots are the other kind. Their seeds only have one seed part so they don't fall into two pieces. Can you guess what "mono" means? It means one. Grasses like Bermuda and corn are monocots. Corn seeds don't have two parts, do they?

So, mono means one and monocots have one seed part. Di means two and dicots have two seed parts. The cotyledon is full of food for the growing baby plant, and if you get there first, for you!

Now go take another look at all the seeds in your kitchen. Will a dried pea split in two? How about a grain of rice? Keep looking and you will be a dicot detective.

Seed Lab #1: INSIDE THE SEED - instructions

**Plan ahead. Seeds must soak overnight.

Material:

copy of lab sheets (2 pages), pencil
1 cup for soaking seeds in
11 or more corn seeds
11 or more dry kidney beans
water
1 flat dish
1 centimeter ruler
crayons - blue and orange

Aloud: A seed is made of the baby plant, called the embryo, a seed coat which surrounds and protects the seed just like your coat does for you on a cold day, and the cotyledons. Cotyledons feed the baby plant until it can make food for itself. Remember, dicots have 2 cotyledons and monocots have 1. Today we're going to pull apart a seed and learn its parts.

Procedure:

The night before:

1. Put 6 corn and 6 bean seeds in the cup.
2. Cover with water and soak overnight.

Lab Day:

1. Place one soaked bean on dish.
2. Remove outer seed coat and check it out.
3. Separate the cotyledons.
4. Investigate seed and label diagram
5. Put one soaked corn kernel on plate. Try to find where this one separates. (It doesn't because it's a monocot). Draw corn seed on lab sheet. Note where it was attached to the cob.
6. Complete graph on lab page 2.

Possible answers:

#6 Seeds soak up water in preparation for sprouting. This makes them bigger than the dry seeds.

#7 Seeds can't sprout as long as they are dry. This is why they don't sprout in the packets at the store. They need the food stored in the cotyledons but they also need the water soaked up from the first rain (or watering). Once the roots have grown, they will provide water for the plant.

For More Lab Fun:

1. Soak an unopened packet of seeds overnight or place a bunch of soaked seeds into a dark container. Place in a warm location and wait. Open after 5 days or so. Did they sprout? Do seeds need sunlight to sprout? Would they receive sunlight underground? Cotyledons provide food energy while the plants are underground. Once they reach the light they can produce their own energy through photosynthesis.
2. CAREFULLY - IODINE STAINS AND IS TOXIC - place a few drops of iodine on the open bean seed. Iodine turns black in the presence of starch, a stored food. Wait a few seconds and see what happens.

NAME ______________________ DATE ______________________

Seed Lab #1: INSIDE THE SEED-pg. 1

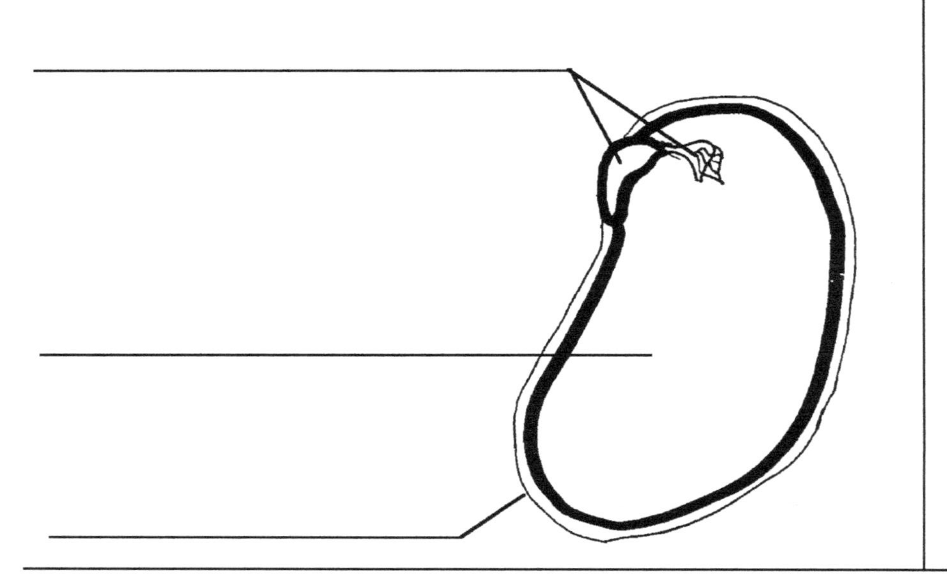

Draw your corn seed here.

Label the cotyledon, seed coat and embryo (baby plant) on the seed diagram.

1. Is your bean a monocot or dicot? (circle one) Is the corn a monocot or dicot?
2. Lay 5 wet beans end to end along your ruler. My wet beans are ________ centimeters long.
3. Lay 5 dry beans end to end along your ruler. My dry beans are ________ centimeters long.
4. My wet corn seeds are ________ centimeters long. My dry corn seeds are ________ centimeters long.
5. How are the wet beans different from the dry beans? ______________________
6. What happened to make the wet seeds different? ______________________
7. What keeps seeds from sprouting in a packet of seeds? ______________________

NAME ______________________ DATE ______________________

Seed Lab #1: INSIDE THE SEED- pg. 2

Color in the graph to show how long, in centimeters, each set of 5 seeds is.
Use a **blue**crayon for the wet seeds. Use an **orange**crayon for the dry seeds.

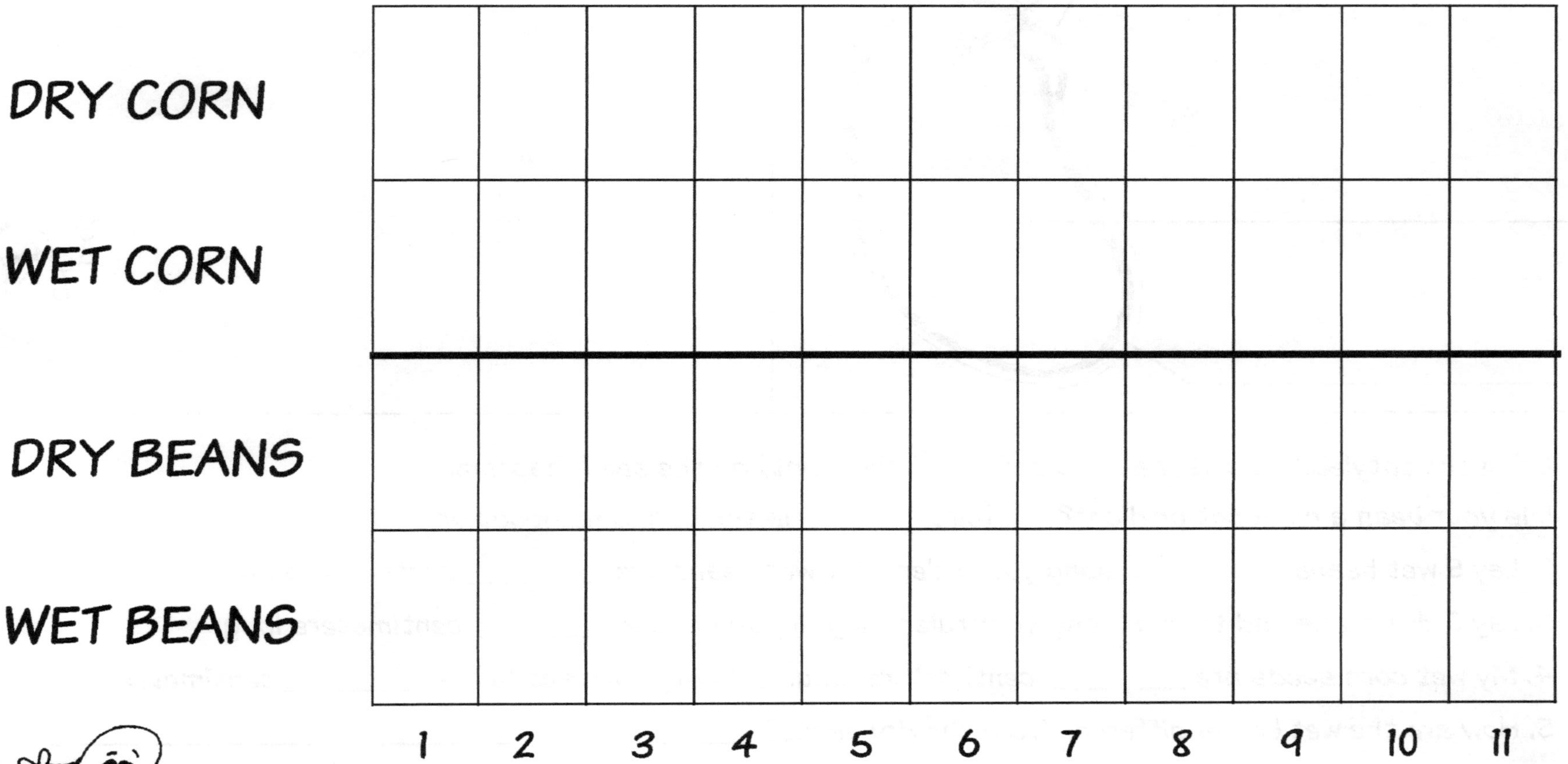

In centimeters, how long are your 5 seeds?

Seed Lab #2: TRAVELING SEEDS - instructions

Material:

copy of lab sheets (2 pages), pencil
big, old socks that you don't mind wasting
hand lens
cranberry or coconut
fruit- any fresh fruit that still has the seeds
seeds that travel by wind- elm, maple, dandelion (kids can help collect these at a park)
tweezers
kitchen knife

Aloud: Sharing one plate of food with all of your relatives would leave you all hungry. The same is true for plants. If all seeds fell right under the parent plant, none of the new plants would get enough food or water or space to grow. Plants handle this by spreading out, away from the parent plant. We will look at four main ways that plants use to get to new areas. Some seeds blow away in the wind, others have special spikes, hooks or sticky glue to grab onto clothing and fur so they will be carried to new areas. Some seeds are carried by water. Another way to travel is to be part of a yummy fruit that gets eaten by an animal and dropped in a new area with the animals drop-pings. We are going to take a close look at how seeds travel.

Procedure:

Lab Day:

1. *Put on a pair of boots and then pull the old socks over the top of the boots. Run, romp and play through a field full of dry weeds. While out, try to find some dry dandelion heads and berries or fruit.*
2. *Take off your outer socks and start picking the stickers out of them. Use tweezers for the "spiky" ones. Look closely and try to pick off at least one of each kind. Place them on the paper plate.*
3. *Use your hand lens to look very closely at each one. Place them on the lab sheet in the space they belong. Which ones have hooks? Which ones are sticky like glue? Which ones form a cork-screw or have barbs that keep the sticker moving into the sock further?*
4. *Now use your hand lens to find the seed that is attached to each sticker. Some may be inside, others very visible.*
5. *Place the cranberry in water. Does it float? Why do you think it might? Have an adult cut the cranberry open. How does the inside look? (It has 4 air spaces to allow the fruit to float).*
6. *On page one, tape down one of each seed in the box that best describes its method of travel. On page two draw the inside of the cranberry (or coconut). On the right of the same circle, draw any fruit that has seeds. First draw the fruit with the seeds showing (usually on an opened fruit), next separate a seed and draw it alone.*
7. *Finish the lab.*

Possible Answers:

#4. Air spaces allow the cranberry to float so it can be transported by water.
#5. A peach has one seed (pit) in it.
#6 Fruits are brightly colored so birds and animals can easily pick them out. They NEED to be eaten so the seeds can be transported to a new location.
#7 To have space to grow, seeds must be able to move to new areas, away from the parent plant.

For More Lab Fun:

1. *With the remaining stickers still attached, lay a sock in a small tray. Cover with potting soil and keep moist. Wait a couple of weeks. Did your traveling seeds sprout?*
2. *Have a "helicopter" versus "parachute" competition. Which type of seed stays in the air the longest or travels the farthest when dropped?*

NAME ______________________________ DATE ____________________

Seed Lab #2: TRAVELING SEEDS- pg. 1

1. Place one of each kind of seed that travels by sticking to something in the correct section of the circle below.

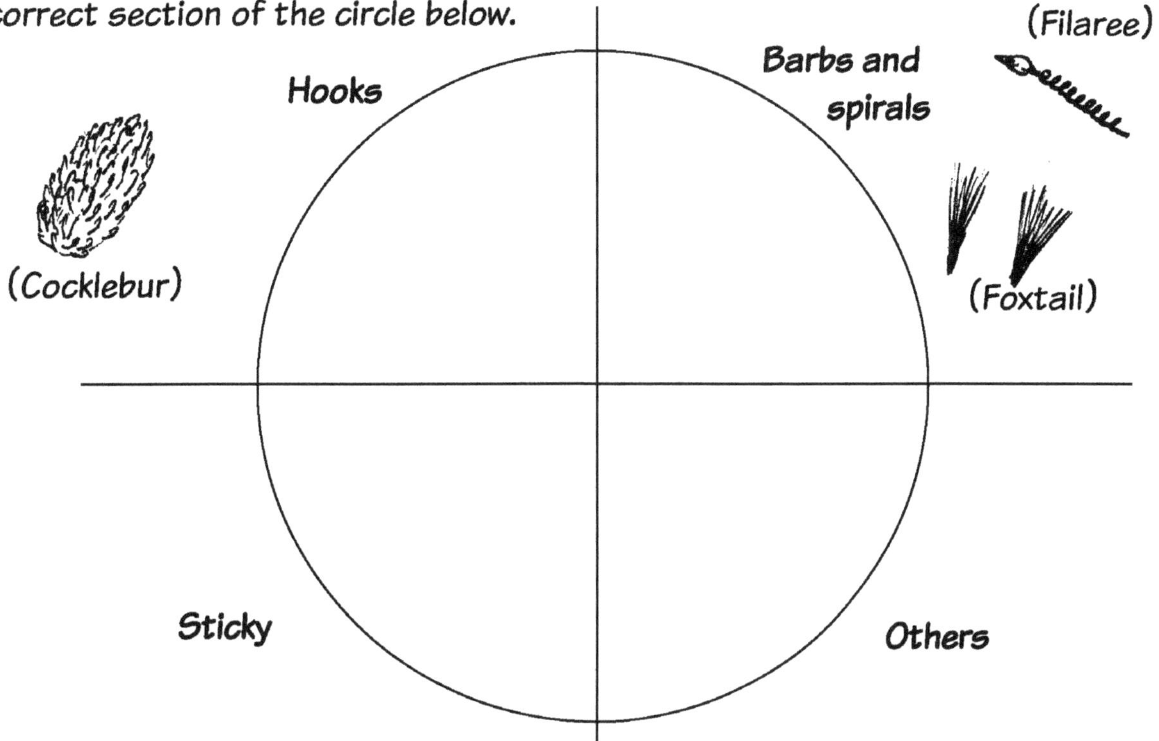

2. Place one of each kind of seed that travels by air in the correct section of the circle below.

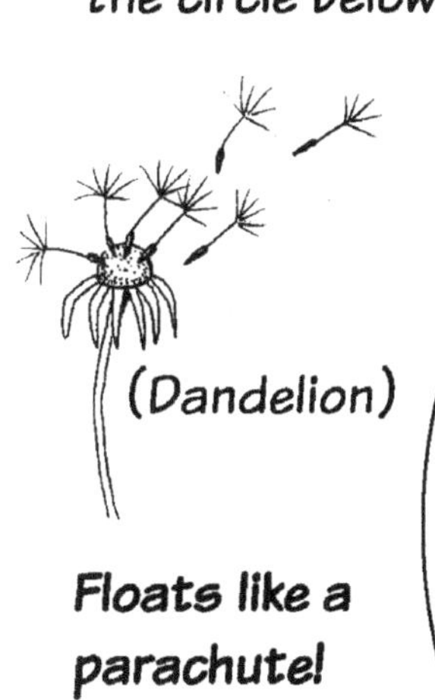

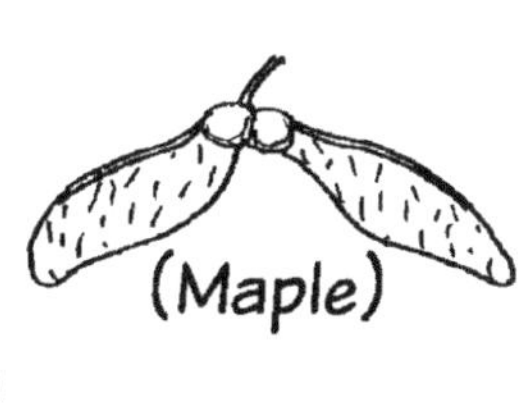

Floats like a parachute!

Spins like a helicopter!

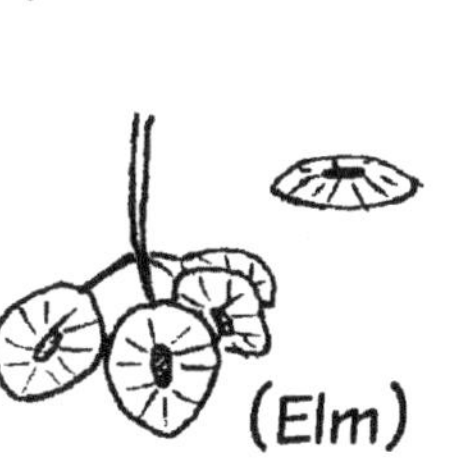

Seed Lab #2: TRAVELING SEEDS- pg. 2

3. Have an adult open a cranberry for you. In the left side of the circle, draw what you see inside. On the right side, draw any fruit, opened, with the seeds inside, and then draw a close-up of one seed.

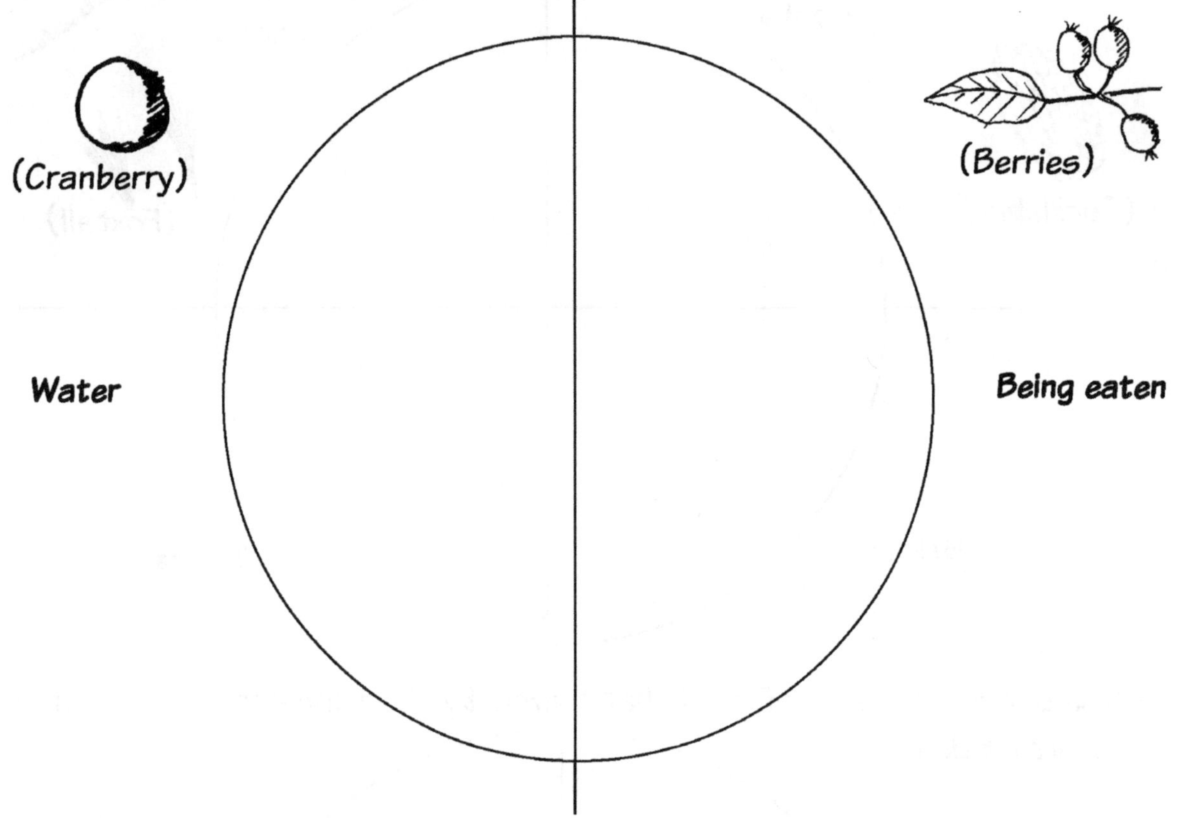

4. What did you notice inside the cranberry that would make it float?

____

5. My fruit has _______ seeds inside. A peach has _____ seeds inside.

6. I think berries and fruit are brightly colored because ________________

____

7. The best way for a seed to have plenty of space to grow is to _______

____

NAME ______________________ DATE ______________

For my notebook

LEAVES MAKE FOOD FOR PLANTS AND ANIMALS

Can a bunch of colored dots change the color of the whole object? They can. A French painter named Seurat (soo-**rah**) made many paintings with nothing but dots of color. Do you remember when we made a model of blood and put little Red Hot candies in it? In blood, a bunch of little red blood cells make the whole blood look red. In leaves, a bunch of green chloroplasts make the whole plant look green. Chloroplasts are green football shaped parts that make food for the plant. Without chloroplasts, a plant would not be green. Of course, without chloroplasts they also wouldn't be alive because they wouldn't be able to make food. Chloroplasts are like cooks. They take ingredients like sunlight, minerals and water, they mix and change things around and make food! Without chloroplasts in the plants even animals would not be able to live. Remember, animals can't make their own food so they have to eat other things. Some animal, like deer, eat plants, some animals, like bobcats, eat animals that ate plants. All food originates in plants which get their ingredients from the sun, soil and air.

So, next time the sun is hot and you wish it would go away, thank it instead for providing the sunlight to make the food for all the living things on earth!

Leaf Activity: WHAT MAKES A LEAF GREEN? - instructions

Material:

computer with internet access (optional)
Sunday cartoon section of the newspaper (optional)
pencil
markers- green, blue, yellow, red and black
3" x 5" index cards white paper

Introduction to Teachers:

This is really a lesson in the art of Georges Seurat, the artist that invented pointillist painting as well as a lesson in the physics of light and mixing of colors and a lesson in the eye and how it blends or "averages" the colors we are seeing. Many objects in nature, that look like solid color are really dots of different colors.

Aloud: Colors can fool us. For instance, blood looks red but it is really pale straw colored with lots of red dots that make the blood look red. Leaves look green but they are really pretty colorless except for the green chloroplasts that are in them. You are going to do an activity today that shows how dots of color can make a whole object look colored.

Procedure:

Lab Day:

1. Go to **http://www.artunframed.com/Seuratthumb.htm** and view some of the art of Seurat. All of his paintings are made with dots of pure colors. Mostly he worked with blue, yellow, red, black, white and orange to make all of the blends you see in his paintings.
2. Look closely at the Sunday comics. Use a hand lens to really see how the comics are made of points of color. Compare this to a plant cell that is mostly a large space but has many tiny chloroplasts that give the entire cell a green color and to the blood model you made. Even though less than half of your blood is made of red blood cells, your blood looks like a red liquid.
3. Many people's eyes are a complex mix of colors. From a distance, what color are your mother's (father's, friend's, teacher's etc.) eyes? Now look closely. Do you see dots of colors that you hadn't seen from a distance?
4. Lightly draw a leaf on one of the index cards with your pencil. With your green marker, fill the leaf with tiny dots of green. Erase the pencil. Look at the leaf from a distance. Can you see how dots of color can appear to fill in a whole object with color?
5. Experiment with colors. Draw a leaf on another card. Fill it with yellow dots then add a number of faint blue dots. View from a distance. What color have you made? Seurat would add dots of orange to give the impression of sparkling sunlight. Try other combinations to see what colors you can make.
6. Draw a picture of your choosing on the paper. Fill it with tiny dots of color like Seurat would have done.

For More Lab Fun:

1. Do a Seurat painting using tempera paints and applying them with the end of cotton swabs—one for each color. Lets the paint dry between colors.
2. Try to use your pure colors to make index cards of pink, orange, brown, and purple.
3. Find a latch hook rug kit or embroidery kit to do that uses dots of color (the yarns) to make new colors.

For Your Information:

Georges Seurat lived between 1859 and 1891.

Leaf Lab: COLOR ME GREEN - instructions

Material:

copy of lab sheet (1 page), pencil
plant with plenty of healthy leaves
variety of color crayons (greens and yellows)
foil

Aloud: Plants are green because they have tiny green chloroplasts in each of their many cells. Chloroplasts have a chemical in them that turns green in the sunlight. So what color are the chloroplasts when they don't get any sun? Let's find out. Today we are going to cover up some leaves on a plant. After a week we will come back and uncover them to find out what happens to leaves that don't get any sunlight.

Procedure:

Lab Day:

1. *Answer the hypothesis on the lab sheet.*
2. *Choose 3 nice green leaves to cover. DON'T PICK THEM! Choose a crayon that closely matches the color of the leaves you are going to cover. Pick one leaf ONLY and place it directly under the top box in the lab sheet. Rub the side of the crayon over the leaf to make a leaf print. It should look very much like the leaves you are covering.*
3. *Gently fold the foil in half around each of the 3 leaves you have chosen. They should be completely covered but not "squished" by the foil.*
3. *Return your plant to the sunlight and water as usual.*
4. *In one week, uncover the leaves and complete the lab sheet, this time choosing a crayon that matches the color the leaves are after a week of being covered.*

Possible Answers:

#2 Most leaves will turn yellow without sunlight.

Conclusion / Discussion:

1. *While the leaves were covered, were they able to make food for the plant?*
2. *What would happen if you tried to grow a plant in the closet?*

For More Lab Fun:

1. *Do an experiment. Buy 2 identical plants. Water them the same but keep one in a dark closet and the other in a nice sunny location. Measure their growth, compare their color and feel the leaves for thickness at the beginning of the experiment and after a week.*
2. *How different would our world look if chloroplasts turned blue in the sunlight? How about orange? What would be your favorite color for a forest? Draw a picture to show this.*

NAME ______________________________ DATE ________________

Leaf Lab: COLOR ME GREEN

HYPOTHESIS:

1. I think the covered leaves will turn __________________ in a week.

Leaf before

Leaf after

2. The covered leaves turned ________________. This is the color of chloroplasts that don't get sunlight.

NAME ______________________________ DATE ____________________

For my notebook

STEMS AND ROOTS DELIVER FOOD AND WATER

We have learned about many parts of a plant. We have learned that flowers attract animals to pollinate them so they can make seeds and that a seed contains the embryo to start a new plant. We also learned that leaves make food from the sun, but why do plants have stems and roots? Roots hold a plant steady so it doesn't fall over every time the wind blows. It takes a strong set of roots to hold up a large tree. Also, roots are very important to a plant because they soak up water from the soil. All living things need water, don't they? This water, along with important minerals travels up into the roots and then through the stem to the leaves. The food made in the leaves also travels through the stem to all parts of the plant. There are special tubes that run through the stem. Food travels in one set of tubes and water travels in another set. With a simple experiment you will be able to see these tubes yourself and see how water goes up the stem into the leaves and flowers.

So, roots grow into the soil to hold a plant up and they also soak up water from the ground.

Stems have special tubes to carry water up from the ground and different tubes to carry food throughout the plant.

Stem / Root Lab #1: WATER PLEASE! - instructions

Material:

- copy of lab sheets (2 pages), pencil
- 2 identical, small, upright, potted plants (something that needs regular watering so you can see results quickly) tomato or bell pepper plants work well
- foil
- spray mister with water
- centimeter ruler

Aloud: If you were thirsty, would you quench your thirst by pouring water on your feet? Probably not. A plant can't take in water just anywhere either. Today we are going to set up an experiment to find out where water enters the plant. It will take a few days to see what happens but if we water the wrong part of the plant, the plant should get thirsty and start to look wilted. This shows us that water must enter the right part of a plant, just like it must enter the right part of you when you are thirsty.

Procedure:

Lab Day:

1. Do the Hypothesis part of the lab.
2. On one pot, write" Plant #1- Roots." On the second pot write "Plant #2- Leaves."
3. With the foil, form a skirt around the base of plant #2 that extends beyond the top edge of the pot. Form into place around the stem (but not around the pot). You are trying to keep water from dripping down into the soil.
4. Answer lab questions 4 and 5. Measure from the table to the top of the plant as is stands. Don't stretch the leaves up.
5. Water the unwrapped plant (Plant #1) by putting the water on the soil, like you would normally do. Water the wrapped plant by misting the leaves. Leave both plants in a good, sunny location. Water them every couple of days, as you normally would. Make sure to water them both at the same times and with about the same amount of water.
6. After a week or two (as soon as you see a plant start to wither) finish the lab. Compare how much each plant grew (or shrank) and how thick or wilted the leaves feel.
7. For #8 and #9, subtract the original height of the plants from the ending height. You may have to do that backwards (A - B = C) if the wilted plant shrunk!

Possible Answers:

1. Plant #1 should have grown and should look healthy. Its leaves should feel thick and strong compared to plant #2. Plant #2 shouldn't get any water inside. Water enters the plant at the roots so it should be wilted. It might have shrunk from losing water or probably didn't grow as much as plant #1. Its leaves should also feel wimpy, thin and wilted.

Conclusion:

1. What would happen if you pulled up a plant, breaking off its roots, and tried to plant it somewhere else? How well do you think it would grow?
2. When you perform an experiment, you usually do what is considered normal to one item and do the test on the other. For instance, you might water one plant with water (the normal) and water another plant with mineral water (the test). The normal one is called a "control." Which plant is the control in the experiment you just did?

For More Lab Fun:

1. Set up a plant experiment of your own. Make sure to use one plant as a control.

NAME ______________________ DATE ______________

Stem / Root Lab #1: WATER PLEASE! - pg. 1

HYPOTHESIS (MY BEST GUESS):

1. I think that water enters the plant through its ______________

2. If I water only the stem and leaves, I think the plant will ________

____

3. If I water only the roots, I think the plant will ______________

____

TEST (THE EXPERIMENT):

Before the experiment:

4. Plant #1, that will have its roots watered is ______ cm. tall.

   Its leaves feel (thick, thin, stiff, soft...) ______________

5. Plant #2, that will have its leaves watered is ______ cm. tall.

   Its leaves feel ______________________________

At the end of the experiment:

6. Plant #1 is ______ cm. tall.

   Its leaves feel ______________________________

7. Plant #2 is ______ cm. tall.

   Its leaves feel ______________________________

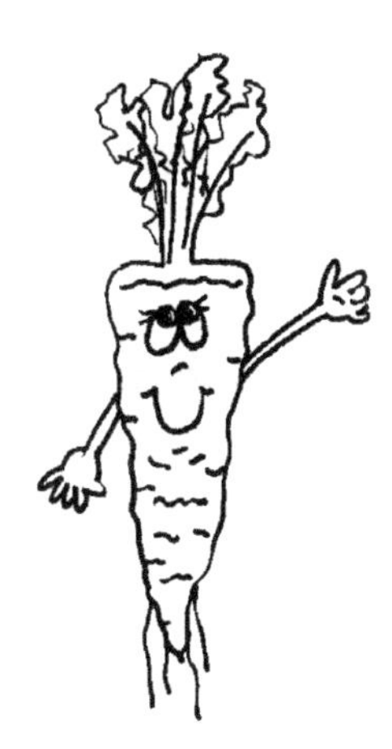

Stem / Root Lab #1: WATER PLEASE! - pg. 2

RESULTS:

(B) Plant height after - (A) Plant height before = (C) Change in Plant

8. Plant #1: ________ cm. - ________ cm. = ________cm.
(B) (A) (C)

9. Plant #2: ________ cm. - ________ cm. = ________cm.
(B) (A) (C)

10. Plant #________ grew the most.

11. The leaves on plant # _______ look the healthiest.

12. Plant # _____ looks the healthiest now because ______________

____. It got

water through its __________________.

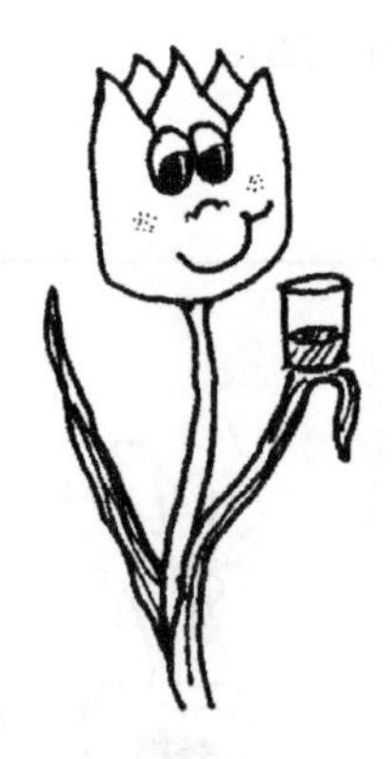

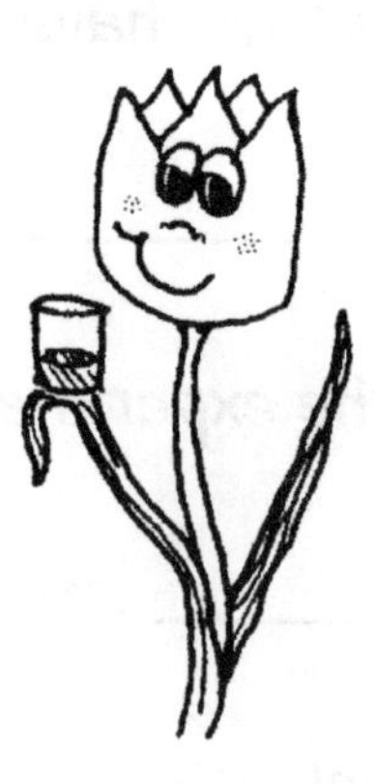

MY CONCLUSION:

I think water enters a plant through its ____________ because . . .

____

____

Stem / Root Lab #2: STEMS MOVE WATER - instructions

Material:

copy of lab sheets (2 pages), pencil
stalk of celery with leaves on top
1 white carnation
3 clear glasses
water
clock
kitchen knife or shears
plate to cut on
food coloring: red and blue colored
pencils: blue, green, and red

Aloud: Roots pull water from out of the ground. This water goes up through the stem in tubes. It's hard to see the water or the tubes when the water is clear so today we will put coloring in the water to see this happen better.

Procedure:

Lab Day:

1. Fill each glass about 1/3 full of water.
2. In two glasses add enough red food coloring to make the water bright red. In the third glass add enough blue food coloring to make the water bright blue.
3. Place one blue and one red glass side by side.
4. Cut about 1/2" off the bottom of the celery stem and the flower stem. Cut the carnation stem in half down the length of the stem as shown in the diagram on the lab sheet. Place the celery in one cup of red water and place the flower with one leg into each color.
5. For questions 1 and 3 have children note time stems went into the water. Finish lab sheet page 1. Leave this set up for 24 hours, but allow kids to check for changes whenever they want throughout the day. When they see color appear in the flowers or leaves, have them mark the time on lab sheet #2 questions 6 and 9.
6. Slice off a small section of the celery. Have children draw what they see in the celery slice diagram on page 2. They should be able to see food coloring in the tubes (xylem) where water travels up the stem.

Possible answers:

#7 and 10- help children calculate how much time it took for the colored water to show up in the stem of the plant. Using a clock with movable hands works well with very small children.
#11. Many answers are fine here. In tubes, up the stem, from the roots . . .

Conclusion / Discussion:

1. Are the two colors in the flower mixed all over or is one half blue and one half red?
2. Does this mean the tubes go all over or straight up?
3. Water is being sucked up the plant. Put 4 straws together and try to suck up water from a glass. Is it harder than you thought it would be? How many straws does it take before it becomes difficult? Think about how hard a tall tree must pull for the water to go all the way to the top leaves!

For your information:

Xylem (zie-lum) is the name for the tubes that take water up into the plant. Phloem (floem) tubes transport food from the leaves throughout the plant.

NAME ______________________________ DATE ____________________

Stem / Root Lab #2: STEMS MOVE WATER-pg. 1

When I set up the experiment, this is how it looked: (Color below)

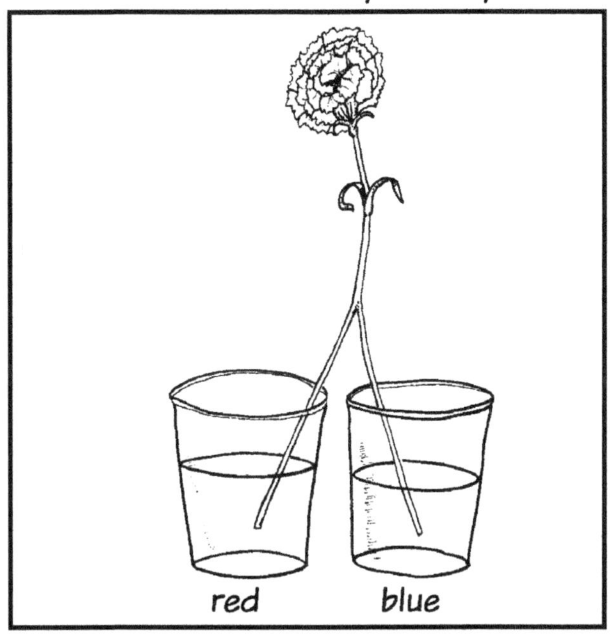

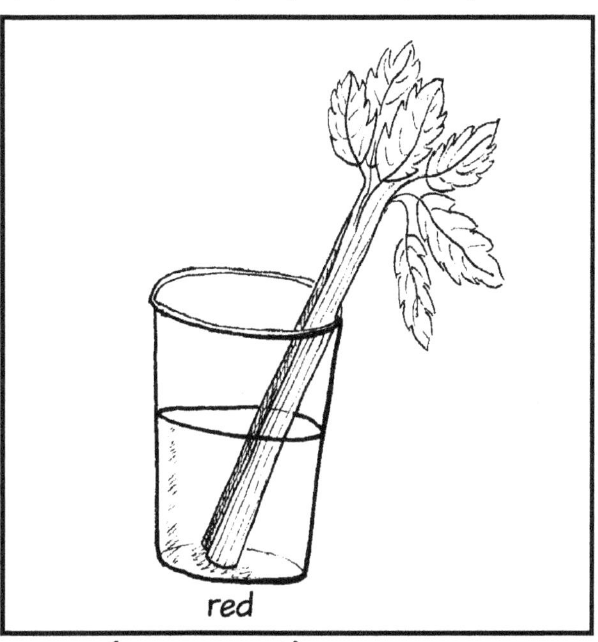

1. I put the flower in the water at _____:______ (What time?)
2. I think it will take _______________ to see a change (How long?)

3. I put the celery in the water at _____:______
4. I think it will take _______________ to see a change

This is what I think will happen: (Color below)

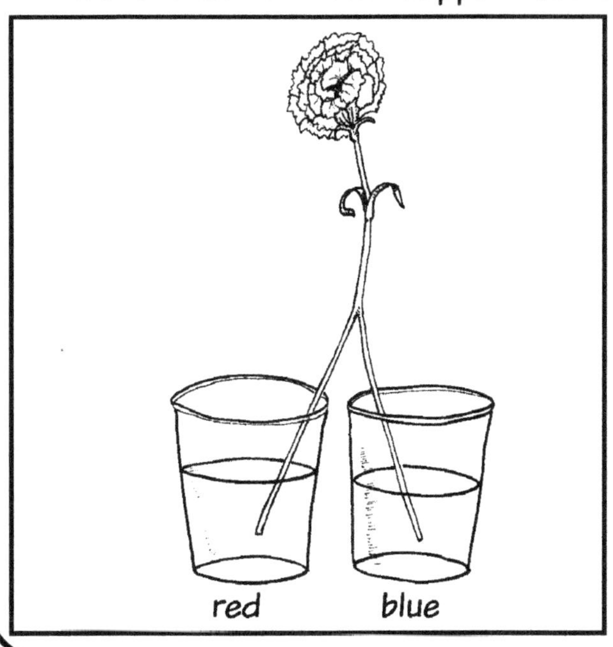

Stem / Root Lab #2: STEMS MOVE WATER-pg. 2

This is what happened with the color: (Color below)

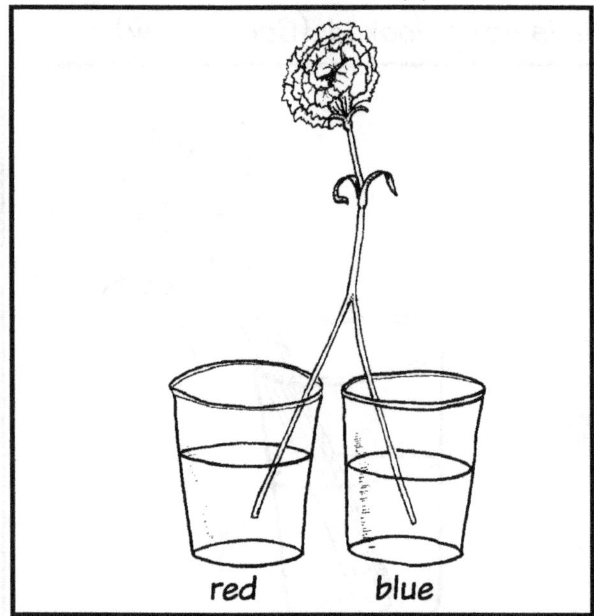

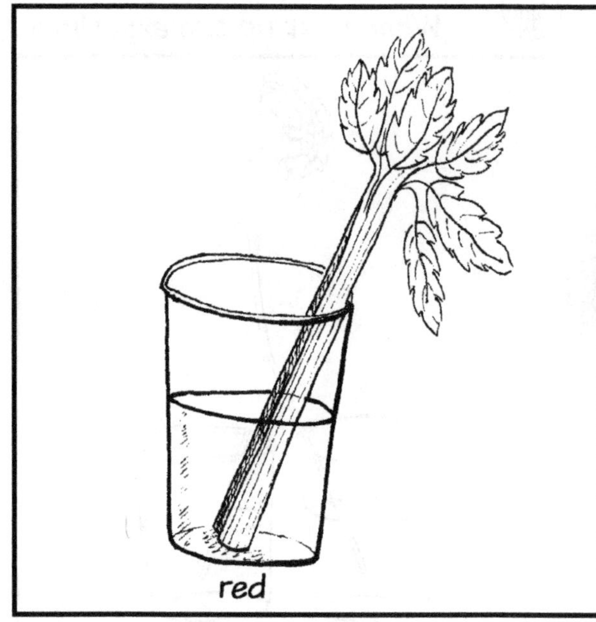

On the diagrams, draw an arrow to show which way the water went.

5. I put the flower in the water at _____:______ (What time?)
6. I first saw color in the flower at _____:______ (What time?)
7. It took ______________ (how long?) for the color to get to the top of the flower.
8. I put the celery in the water at _____:______
9. I first saw color in the celery at _____:______ (What time?)
10. It took ______________(how long?) for the color to reach the top of the celery.

Color in the flower close up

Color in the end of the celery

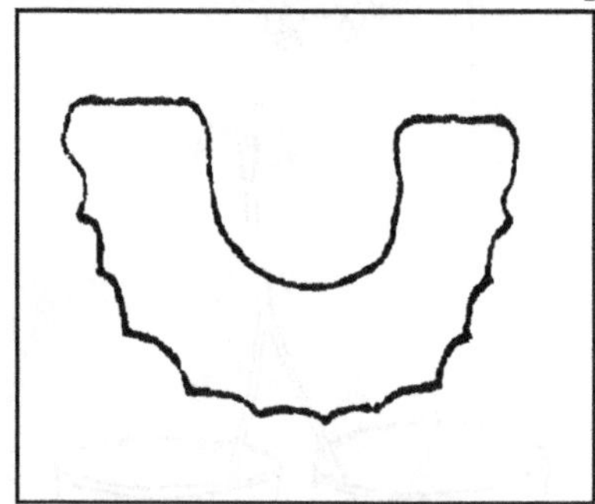

11. I can see that water travels through a plant____________________

____

Plant Summary: PLANT PARTS SALAD - instructions

Material:
- copy of lab sheet (1 page), pencil
- kitchen knife
- salad bowl
- salad dressing of choice
- whatever you make a salad from!

Aloud: Now that you know about the different parts of the plant- flower, seed, stem, leaves and roots, it's time to put all that together into a salad! Think of what you put into a salad. What part of the plant are you eating? Is it a leaf, stem, flower? Write it down in the category it belongs. What else do you put in a salad? Keep going until you have a few types of food in each category. Think hard. I bet you eat food from every part of a plant. When you are done, plan a trip to the store. Take your paper so you can write in more foods you hadn't thought of. Buy what you like in a salad, trying to get something from each part of the plant. You will then bring them home, wash and cut them, add your favorite salad dressing and enjoy a plant part salad! Bet you never looked at broccoli or sunflower seeds that closely before.

Procedure:

Lab Day:

1. Brainstorm different salad fixings and decide where they come from on a plant. Write them down on the lab paper where they belong.
2. Make a shopping list and buy salad parts from each part of the plant.
3. Wash, chop and make a salad with the parts you have chosen. Eat up, but not without thinking about where your food has come from!

Possible Answers:

Suggestions for plant parts list:

Flower: broccoli, cauliflower, artichoke

Leaves: lettuce, spinach, cabbage

Stem: celery, rhubarb, asparagus

Roots: carrot, onion, beet

Seeds: sunflower seeds, beans, peas

NAME ______________________________ DATE ____________________

Plant Summary: PLANT PARTS SALAD

Make a list below of the foods you might put in your salad. List each one according to the part of the plant it comes from. Make sure to include items from each part of the plant. Take this list to the grocery store with you and add more as you shop. When you get home, put them together and have a salad!

FLOWERS

____

____

____

LEAVES

____

____

____

STEMS

____

____

____

ROOTS

____

____

____

SEEDS

____

____

____

SCIENCE VOCABULARY

Abdomen- On an arthropod, the furthest body segment from the head. On a human, the area between the chest and hips.

Amphibian- Cold blooded, vertebrate animal that starts life with gills, living in water and develops lungs to breathe air as it mature. Frogs, toads, salamanders.

Anemone- Animal of the phylum Cnidaria. Anemones are stationary, mostly stuck to rocks on the sea floor. They catch food with their stinging tentacles as it floats or swims by.

Antenna- Long feelers on the heads of insects and crustaceans. (Plural: antennae)

Anus- End of the digestive tract where solid waste leaves the body.

Arachnid- Group of arthropods with eight legs, two body segments, no antennae and no wings.

Arteries- Blood vessels that carry blood away from the heart, to all parts of the body.

Arthropod- Group of animals with jointed legs and segmented bodies. Insects, spiders, crabs...

Blastodisc- Tiny white spot on the outside of the egg yolk. It contains most of the cell parts, including the nucleus and along with the yolk is one cell.

Blood- The fluid that is pushed through the circulatory system by the heart. It carries oxygen, food and waste.

Blood vessels- The tubes in the body that blood circulates through. Arteries, capillaries, veins.

Brain- The major organ of the nervous system. Controls memory, emotions, movement...

Bug- Member of a group of insects with piercing, sucking mouth parts.

Carbon dioxide- Gas made of oxygen and carbon. Produced in animals by breathing.

Cell- The most basic building block of all living things. Some living things are made of only one cell. Most are made of a huge number of them.

Chalaza- Rope-like strands that anchor the egg yolk in place

Chloroplasts- Cell parts inside plant cells that contain chlorophyl, a substance that converts sunlight into food.

Chrysalis- Pupal stage of a butterfly or moth when it appears inactive and out of which will emerge the adult butterfly or moth.

Circulation- Movement of blood or another substance through an organism.

Circulatory system- The organs and structures that move blood through the body.

Clavicle- (aka collarbone) The long, thin bone that runs from each shoulder to the base of the neck.

Clitellum- (aka cuff) Raised section on a mature earthworm's body, used for reproduction.

Cnidaria (aka Coelenterate)- Primitive group of animals with stinging tentacles, a hollow body and radial symmetry. Corals, anemones, sea jellies.

Cold blooded- Describes animals whose body temperatures change with the temperature of its surroundings. All invertebrates as well as fish, amphibians and reptiles.

Coral- Tiny, warm water animals in the phylum Cnidaria who live in colonies and produce masses of hard protective exoskeletons which, together form reefs.

Cotyledon- (aka seed leaf) Sections of a seed where food is stored for the emerging embryo.

Crustacean- Group of arthropods with ten or more appendages, no wings, a hard exoskeleton and two pairs of antennae.

Dicot- (aka dicotyledon) Any flowering plant with two seed parts and spreading veins within the leaves.

Digest- To break down foods into chemicals that can be used by the body.

SCIENCE VOCABULARY

Digestive system- *The organs and structures of the body that are involved in the process of breaking down foods into chemicals for the body to use.*

Dissect- *To cut something apart in order to see its inside structure.*

Dormant- *Not active for a time.*

Echinoderm- *Phylum of animals with spiny skin and a body plan that, in someway involves five parts.*

Embryo- *An animal or plant at a very early stage of development.*

Esophagus- *The tube that carries food from the mouth to the stomach.*

Exoskeleton- *An external, protective structure of many invertebrate animals such as arthropods.*

Femur- *(aka thigh bone) Long bone in the upper leg. The longest bone in the human body.*

Fertilize- *To provide sperm or pollen to an ovum in order to make it capable of producing offspring.*

Fin- *An appendage (like an arm or leg) shaped for swimming.*

Gene- *A tiny part of a living thing's cells that determine a characteristic that living thing will have and can pass on to it's offspring*

Gill- *A breathing organ of some animals that takes oxygen from the water.*

Gill cover- *A protective flap of tissue covering the gills.*

Heart- *The major organ of the circulatory system, it is a muscular organ that forces blood through the body.*

Humerus- *Long bone in the upper arm, running from the shoulder to the elbow.*

Inherit- *To receive a characteristic through genetic transmission from a parent.*

Intestine- *The part of the digestive system below the stomach where food and water are further digested and absorbed.*

Invertebrate- *An animal without vertebrae (backbones).*

Isopod- *A member of the group of crustaceans which are commonly called sow bugs, pill bugs or roly-polies.*

Kingdom- *First level of division of all living things. Most scientists agree on splitting all living things into 5 kingdoms: Monera, Protista, Fungi, Plant and Animal.*

Larva- *The immature stage of certain animals when they often appear very unlike the adult. A caterpillar and a tadpole are larvae. (Plural- larvae)*

Lung- *Major organ of the respiratory system. These spongy organs take in air, extract the oxygen and gather carbon dioxide from the body to be released back into the air.*

Mammal- *One of a group of vertebrates with hair, that do not lay eggs and that feed milk to their young.*

Mandible- *(aka jawbone) The lower jawbone.*

Medusa- *The tentacled, bell shaped, free swimming form of some Cnidaria such as sea jellies.*

Metamorphosis- *A complete change in form or appearance as the one that occurs in butterflies. Complete metamorphosis occurs in four stages: egg - larva - pupa - adult. Incomplete metamorphosis occurs in three stages: egg - larva - adult.*

Mollusk- *The phylum characterized by a soft body, complex organs and usually a hard shell. Clams, octopi, nautilus, snails...*

Monocot- *(aka monocotyledon) Any of the flowering plants with one seed part and parallel veins within their leaves.*

SCIENCE VOCABULARY

Muscular system- The group of body tissues whose function is the movement of body parts.

Nerve- One of the bundles of long, thin cells that carries messages between the spinal cord or brain and the parts of the body.

Nervous system- The brain, spinal cord, nerves and sense organs all working together to gather information and to act upon it by controlling all body functions and movements.

Nucleus- Major part of most cells that controls all aspects of the cell, including heredity.

Organs- A group of body tissues formed together to perform a common function. Heart, lung...

Ovule- A tiny plant structure which, after fertilization becomes a seed.

Oxygen- Colorless, odorless gas required for life by most living things.

Patella- (aka kneecap) Disk-like bone at the knee, covering the joint between the lower and upper leg bones.

Pelvis- (aka hip bones) The flat, platelike set of bones at the lower end of the abdomen.

Petal- A brightly colored, leaf-like part of a flower whose function is to attract pollinators.

Phalanges- Small bones of the fingers and toes.

Phylum- A division in the groupings of living things. Next specific level following "kingdom".

Pistil- The female part of a flower where seeds develop. Composed of the stigma, style and ovary.

Plasma- The clear, straw colored liquid part of blood whose function is to transport cells, dissolved food and waste.

Platelet- A tiny, flat disk within the blood whose function is to stop the bleeding when a vessel is cut.

Plot Study- A scientific analysis of a measured area of land.

Pollen- Fine, dust-like material produced by the anther of a flower to fertilize the female cells.

Pollinate- To transfer pollen from the anther of a flower to the stigma.

Pollinator- Any force, living or not that transfers pollen from the anther of the flower to the stigma. Often wind, insects, birds and bats.

Pollen grains- The tiny parts that make up pollen, or the male, fertilizing substance of a plant.

Polyp- The life stage of a cnidaria in which it is shaped like a tube with one end attached to the ocean floor and the other end waving its stinging tentacles in order to gather food.

Pupa- The stage of complete metamorphosis between larva and adult. This is the immobile stage, called a chrysalis in butterflies.

Radula- In mollusks, the flexible tongue covered with sharp hooks for grating food.

Recessive- A gene or trait that is weak or masked by the opposing trait. For instance, the gene for blue eyes is masked when the bearer also has the gene for brown eyes.

Red blood cell- The disk-like cell in the blood that carries oxygen to the tissues and carries carbon dioxide away.

Reproduction- The process by which living things produce offspring or more of their own kind.

Reptile- One of the group of vertebrates with scaly skin and lungs, most of which lay eggs.

Respiration- The process in living things of exchanging gasses. In animals, oxygen is brought in and exchanged for carbon dioxide, which is then removed.

Rib cage- The series of paired bones in the chest that surround and protect the lungs and heart.

Sand dollar- A thin, circular echinoderm with soft spines.

SCIENCE VOCABULARY

Scale- *One of several flat plates that make up the outer covering of fish and reptiles.*

Scapula- *(aka shoulder blade) One of two flat, triangular bones that form the back part of the shoulder.*

Scientific name- *A unique, specific name given to each individual, living species to identify it. The scientific name is made up of the genus and species names. People are Homo sapiens.*

Sea cucumber- *A soft, fat, wormlike echinoderm with tentacles that surround the mouth.*

Sea urchin- *An echinoderm with a soft body surrounded by a round shell covered with long spines.*

Seed coat- *A thin, protective covering around a plant seed.*

Senses- *The way in which a living thing gathers information about what is happening outside its body. Humans have five senses: smell, taste, sight, hearing and touch.*

Sepal- *The green, leaf-like structure that protects the flower while it is still a bud.*

Seta- *A stiff hair or bristle as those on an earthworm. (Plural: setae)*

Simple animal- *An animals with no highly developed organs or organ systems.*

Species- *The most specific division in the classification of living things. Members of the same species are those that can produce viable offspring or, in other words, members that can produce offspring that can then produce their own offspring. A horse and a donkey are different species as they can together produce offspring (a mule) but mules are infertile and cannot produce offspring of their own. Humans are of the species Homo sapiens.*

Spinal cord- *A thick bundle of nerve cells that runs along through the vertebrae connecting the brain to the nerve cells in the arms, legs and trunk of the body.*

Spiracle- *A breathing hole such as the ones that run along each side of an insect or spider or the blowhole of a whale or dolphin.*

Stamen- *The male part of a flower, composed of the filament and anther, where pollen is produced.*

Sternum- *(aka breastbone) The bony plate in the chest that connects the ribs and the clavicle.*

Thorax- *The middle body segment of insects where the wings and legs attach.*

Trachea- *(aka windpipe) The air passage that connects the mouth and nose to the lungs.*

Trait- *A specific feature or characteristic of living things that is inherited from a parent.*

Veins- *A vessel of the circulatory system that carries blood towards the heart.*

Vertebra- *(aka backbone) One of the donut shaped bones of the spinal column. (Plural: vertebrae)*

Vertebrate- *An animal with a spinal column or backbones.*

Warm blooded- *Those animals whose temperatures stay about the same regardless of the temperature of their surroundings. Mammals and birds are the only warm blooded animals.*

Worm- *An animal from one of 3 phyla characterized by a long, threadlike body shape, no legs and fairly complex organs.*

White blood cells- *Cells in the blood that fight invading bacteria and other intruding substances.*

WHERE I LOOKED STUFF UP

Akers, Art et al. 1986. *From Head to Toe,* AIMS Education Foundation, Fresno, California.

Brown, Philip R. 1994. *Exploring Tidepools.* EZ Nature Books. San Luis Obispo, California.

Hoover, Evalyn and Sheryl Mercier. 1990. *Primarily Plants.* AIMS Education Foundation, Fresno, California.

Kneidel, Sally. 1993. *Creepy Crawlies and The Scientific Method.* Fulcrum Resources, Golden, Colorado

Mitchell, Andrew. 1982. *The Young Naturalist.* Usborne Publishing, London, England

Rillero, Peter Ph.D. et al. 1999. *Science Projects & Activities.* Publications International, Ltd. Lincolnwood, Illinois

Unwin, Mike. 1992. *Science With Plants.* Usborne Publishing Ltd., London, England

VanCleave, Janice. 1993. *Animals.* John Wiley & Sons, Inc., New York, N.Y.

Watts, Lisa. 1994. *Usborne Illustrated Encyclopedia of The Natural World.* Usborne Publishing Limited, Saffron Hill, London, England